The Grounde of Artes

Robert Recorde
The Grounde of Artes

A FACSIMILE OF THE
FIRST EDITION
IMPRINTED AT LONDON BY
REYNOLD WOLFE
1543

RENASCENT BOOKS

The Grounde of Artes
First imprinted by Reynold Wolf in 1543
Subsequent Editions
Imprinted by Reynold Wolf
1545, 1549, 1551, 1552, 1558, 1561, 1566, 1570, 1571, 1573
Imprinted by Henry Binneman & Iohn Harison
1575, 1579, 1582
Imprinted by T. Dawson for Iohn Harison
1594
Imprinted by Richard Field for Iohn Harison
1596
Imprinted by N. Oakes for Iohn Harison
1607
Imprinted for Iohn Harison
1610
Printed by Thomas Snodham for Roger Iackson
1615
Printed by Iohn Beale for Roger Iackson
1618, 1623
Printed by Tho. Harper for Iohn Harison
1631, 1632, 1636
Printed by J. Raworth for John Harison
1640
Printed by Miles Flesher for John Harison
1646, 1648
Printed by James Flesher
1652, 1654, 1658, 1662, 1668
Printed by E. Flesher
1673
Printed by J. H.
1699

Hardback facsimile edition published by
TGR Renascent Books
27 Springdale Court
Mickleover, Derby DE3 9SW
United Kingdom
2009

Paperback edition first published 2012

All rights reserved. No part of this publication may be reproduced, stored in a retrieval system, or transmitted, in any form or by any means, without the prior permission in writing of the publisher, or as expressly permitted by law.

© Thomas Gordon Roberts 2009

ISBN 978-1-4810896-0-9
www.renascentbooks.co.uk

Printed and bound by
CreateSpace, Charleston, South Carolina, U.S.A.

for
ELIZABETH

INTRODUCTION

This book is a facsimile of the first edition of Robert Recorde's *The Grounde of Artes*, originally printed in London by Reynold Wolfe in 1543. Recorde was a typical Renaissance man, a physician at the courts of Henry VIII, Edward VI and Mary, a mathematician and a very learned scholar. His book was the first in a series of justly famed mathematical works with the following titles.

> *The Grounde of Artes*
> *The Pathway to Knowledge*
> *The Gate of Knowledge*
> *The Castle of Knowledge*
> *The Treasure of Knowledge*
> *The Whetstone of Witte*

Of these six, *The Gate* and *The Treasure* are no longer extant. The first book in the series (the one you now hold) deals with arithmetic; the later books cover geometry, astronomy and algebra. *The Grounde* was not the earliest arithmetic text published in England. That distinction belongs to a work written in Latin and printed in 1522; Cuthbert Tonstall's *De Arte Supputandi*. Nor was it the earliest book of arithmetic written in English. Such a book was published at St. Albans in 1536; a curious work with the intriguing title *An Introduction for to lerne to reken with the pen and with the counters, after the true cast of arismetyke or awgrym in hole numbers, and also in broken*. But Recorde's book proved to be unprecedentedly popular; it was what today we would call a runaway best-seller. It is probable that everyone who studied some aspect of mathematics in Elizabethan England would have done so by joining Recorde for at least some of the way along the metaphorical journey suggested by the ordered progression of his titles. Many readers, in an age of great voyages of discovery and enterprising seamen, would no doubt have been young men with aspirations to serve as

Introduction

gunners or navigators aboard the country's ships of war and trade – positions requiring no small amount of mathematical skill. The book is written in the form of a dialogue between a master and a somewhat precocious scholar, a format which allows the scholar to make many mistakes before being corrected by the wise but kindly master. It is also that rare thing, a mathematical textbook with humour, as the discerning modern reader will soon realise. There were probably more than forty printings of *The Grounde* over a period of 150 years, the last being in 1699. Today, an original first edition is an extremely rare book, although later 17th century editions are occasionally offered for auction or sale by antiquarian book sellers at prices running into many hundreds of pounds.

BRIEF BIOGRAPHY

Robert Recorde was born between 1510 and 1512 in Tenby, Pembrokeshire. He was the second son of Thomas Recorde and Rose Jones. Practically nothing is known of his childhood and the first thing in his life about which we can be certain is his entrance into Oxford University in about 1525. It is not known what he studied but we may assume it was the usual (for the time) *trivium* of grammar, logic and rhetoric, followed by the *quadrivium* of arithmetic, geometry, music and astronomy. He graduated with a B.A. in 1531 and was elected a Fellow of All Souls College in the same year. He may have taught at Oxford for a few years but the evidence for this is scanty. At some time he moved from Oxford to Cambridge, where he studied for an M.D. and graduated in 1545 at the age of 35. He then moved to London, where for a few years he practised medicine. In later years he was always to describe himself as 'physician'. A defining moment in his life occurred in 1549 when he was appointed Controller of the Bristol Mint. It was during his time there that he made a very powerful and ruthless enemy. Sir William Herbert was sent

Introduction

by Edward VI to help suppress a revolt by John Dudley, Earl of Warwick, in the west country. Herbert demanded that Recorde divert funds from the mint to pay and support his army, but Recorde refused on the grounds that the order did not come from the king. Herbert countered and accused Recorde of treason. He was lucky to incur the mild penalty of confinement to court for 60 days. However, apparently all was later forgiven because in 1551 he was appointed general surveyor of Mines and Monies in Ireland. He was placed in charge of the Wexford silver mines and also became the technical supervisor of the Dublin mint. In the meantime, Sir William Herbert was created Earl of Pembroke for his services to the crown during the rebellion, and there was continued animosity between him and Recorde. Although the silver mines at Wexford had great potential, the enterprise was largely unsuccessful, mainly due to a lack of royal investment and the imperfect state of mining technology. The mines closed in 1553 and Recorde was recalled to England. Upon the accession to the throne of Mary, the daughter of Henry VIII, Recorde's old enemy the Earl of Pembroke was made a privy councillor for his support of Mary's claim to the throne. For some strange reason, Recorde chose the moment when Pembroke was strongest to try and get his revenge, charging him with misconduct in gaining his court positions. The allegation was probably true, but Pembroke was in favour with the monarchy and so had almost perfect immunity. He responded by suing Recorde for libel. There was a hearing in January 1557 and Recorde was ordered to pay the huge sum of £1000 compensation. He either could not or would not pay and so was sentenced to imprisonment in the King's Bench Prison in Southwark, for debt. Whilst in prison he made his will, leaving small sums of money to various people, including £20 to his mother. The date of his death is not

Introduction

known with any certainty, but is generally supposed to have been in the later part of 1558, only a short time after making his will.

The following notes are provided as a guide to reading, understanding and enjoying this facsimile edition of Recorde's treatise.

PAGINATION

The pages are not numbered individually as in modern practice, rather each leaf only is numbered and pages are referred to as side 1 or side 2 of the leaf (see the errata at the end of the book for an example, where pages are listed by Lefe (leaf), fyde (side) and lyne (line). However, the original compositors have confused the leaf numbering. Ordering starts at 1 and proceeds to 40, on the next leaf the numbering restarts at 35 and continues to 122, whereupon the next leaf restarts again at 116 and continues to 137 (which leaf should actually be 150 if numbered consecutively). Caution is therefore required when page referencing, since the errors of the sixteenth century compositors are faithfully reproduced in this facsimile reprint. However, for the benefit of modern day readers who expect consistent pagination, modern page numbering is applied over a line drawn above Recorde's text.

SPELLINGS AND PUNCTUATION

Dictionaries and standard spellings did not exist when *The Grounde of Artes* was first printed and many words are not even spelt phonetically. Therefore many spellings in the book appear peculiar to the modern reader, but a little practice at reading Early Modern English soon renders the text intelligible. A difficult looking word in the text such as fowldyours, if read sowld(sold)-yours is readily deciphered from the context as soldiers. Many familiar words look strange simply because,

Introduction

unlike modern spellings, they end with the silent letter e and the last consonant might or might not be doubled, hence mane or manne (man), and rune or runne (run). The letter y is often used in place of i, for example fynde (find) or fyrste (first). Note that early printing conventions were to use the terminal letter s at the end of words, as today, but the long form everywhere else, for example poſſeſs (possess). The letters u and v were not considered to be two distinct letters, but different forms of the same letter. Typographically, v was often used at the start of words and u elsewhere, hence vnmoued (unmoved) or vnloued (unloved). But conversely, the letter v was often used where today we would expect the letter u, as in, for example, thervnto (thereunto). Neither were the letters i and j considered distinct, so that the name John would appear spelt as Iohn. In short, expect to read modern spellings such as sum, divisor and just for example, as ſumme, diuiſor and iust. Punctuation is eccentric to modern eyes – full stops and the colon are used apparently interchangeably and words following these symbols might or might not be capitalised.

CONTRACTIONS

Almost universally, the words 'the' and 'that' are contracted to y^e and y^t respectively, with the small letters e and t placed directly above the y. Of course, these words should be read with their full pronunciation – 'the' and 'that'. The word 'with' is often contracted to w^t, again with the t in smaller point directly above the w. Sometimes this contraction is encountered in combination, e.g. w^t + out (w^tout) for 'without', or w^t + draw (w^tdraw) for 'withdraw'. A distinctive symbol which looks something like a letter p (þ) is sometimes used by the compositors of *The Grounde of Artes* to represent 'per', so that 'perceive' for example, might appear spelt phonetically in full as 'perceaue', or contracted as Þceaue. Occasionally, the compositor also uses Þ to represent 'par', as in Þcels (parcels).

DIACRITICAL MARKS

Diacritical marks have been used to abbreviate printed words ever since Gutenberg and early English printers adopted the same conventions that Gutenberg did for Latin texts (which he copied, in turn, from the handwritten texts of medieval scribes). Diacritical marks are used on almost every page of *The Grounde of Artes* to indicate the omission of the consonant m or n where this follows a vowel. The missing letter is indicated by placing the mark (a bar) over the vowel. Instances are exāple (example), quotiēt (quotient), ī (in), nōbre for nombre (number) and chaūce for chaunce (chance). All such abbreviated words with diacritical marks should be read of course with their full pronunciation.

TYPOGRAPHICAL FEATURES

All the contractions and abbreviations found in the pages of *The Grounde of Artes* are compositors tricks to help in the justification of entire paragraphs – something that was considerably easier in the days before standard spellings and orthography. Justification of paragraphs was not then merely a cosmetic feature (as it is today). Early printers would be laying out movable metal type into a square wooden frame and if the frame wasn't completely filled, the types would fall out. In other words, each line of each paragraph had to extend fully from left to right, or the page would be unprintable. One way for Renaissance printers to do this might have been by inserting blank spaces of suitable lengths between the words of each line, but this is not a satisfactory solution. The result is usually 'rivers' of blank space flowing down the page, which seriously interrupt reading and which was recognised as a problem from the very earliest days of printing. Hence the use

Introduction

of aggressive hyphenation, contracted words, diacritical marks and variant spellings like hed (which has three letters), head (which has four), or hedde (which has five), all very useful when striving to obtain justification and spelling is not a problem. The compositor would use any or all of these tricks at will in order to obtain a solid block of text on each page.

FAULTS

The errata given at the end of *The Grounde of Artes* lists those faults recognised by the author and his printer 450 years ago. However, the perceptive reader will find more faults and errors than those listed and, of course, no attempt has been made to indicate or remedy these in this facsimile reprint. The erratic page numbering has already been mentioned, but it is not unusual to find letters upside down, so that the letter n looks like u and m looks like w (and vice-versa). Similarly, letters are sometimes juxtaposed so that 'maketh' for example, is printed as 'makteh', or 'turne' as 'tunre'. Sometimes spelling mistakes occur despite the relaxed phonetic spelling. An example here is 'mulitiplication', which has a surplus letter i and is obviously wrong. In a number of cases the running head at the top of each page is not consistent in reflecting the content of the page beneath it. More seriously, mathematical mistakes occur here and there – lines of figures in arithmetical examples do not always add, subtract, divide or multiply as the text says they should. Almost certainly this is due to typographical errors made by the compositors. While necessarily literate, these men were often not equally numerate and their mistakes often betray a lack of appreciation for the subtleties of place values. Nor did they have the arithmetical skills to realise that a result was incorrect because of mistakes in picking and typesetting the wrong numbers from their cases of movable type. When readers encounter these typographical and mathematical errors in the book, remem-

Introduction

ber that they occur in the original printing and are faithfully reproduced in this facsimile reprint.

THE TITLE

Recorde's meaning in his title is that arithmetic is the essential bedrock – the ground – from which all subsequent studies of the arts must spring. The arts are not only those of higher mathematics, such as geometry and algebra, but also what today we would call the sciences, such as astronomy and the newly important science of celestial navigation at sea. But Recorde goes further, as he makes plain in his other books. To him, the arts also include what we call crafts. So artisans such as shipwrights, millwrights and clockmakers for example, would only prosper if they had a thorough grounding in the art of mathematics and arithmetic in particular. On the cover of this book, the title is spelt as *The Grounde of Artes*, but a reference to the title page of the facsimile shows the title as *The groūd of artes*. Later editions had titles with variant spellings, such as *The Grovnd of Arts*, *The Grovnde of Artes* or *The Grounde of Artes*. The later is that which is probably the least puzzling to the modern reader, and therefore the most suitable to appear on the cover.

TITLE PAGE WOODCUT

The illustration on the title page of the facsimile is sometimes called "The Quarrel of the Abacists and the Algorists". Abacists were those who defended Roman numerals and calculations done with counters on ruled boards (a sort of abacus), and they are represented on the left of the picture. The two men on the right are algorists, who supported the written calculation methods done with pen and paper, originally invented in Asia (Arabic numerals) and newly introduced into the west. The disagreement lasted for several centuries. Skilled abacists were long regarded in Europe as magicians

Introduction

enjoying almost supernatural powers, and even with the introduction of written arithmetic, multiplication and division usually remained outside the grasp of ordinary people, given the difficult operating techniques then used. This perhaps explains why books on arithmetic such as this one by Robert Recorde, which teaches both calculation with the pen and with the counters, was absolutely necessary. It was not until the end of the eighteenth century that simpler techniques were generalised and brought basic arithmetical operations within the scope of most adults and eventually, even of children.

TITLE PAGE ICONOGRAPHY

On the back wall of the room in which the abacists and the algorists are arguing over a ruled counter and written numerals, are four letters and a date – the date being the year of publication (1543). The letters VDMIE are more mysterious. They are in fact the initial letters of the Latin phrase *Verbum Domini Manet In Eternum* – 'The word of the Lord shall last forever'. The phrase was adopted as a slogan by the Schmalkaldic League, an alliance concluded on 27 February 1531 between several German protestant princes and cities in opposition to the catholic Holy Roman Emperor Charles V, who at this time was attempting to suppress Protestantism and the teachings of Martin Luther. In England, Thomas Cromwell, chief advisor of Henry VIII in all ecclesiastical affairs, wanted Henry to join the League but in this he was unsuccessful. Subsequently, he endeavoured to bring about an alliance between England and the German protestants by arranging Henry's disastrous marriage to Anne of Cleves. VDMIE became the watchword of the Schmalkaldic League and the rallying cry of Protestantism throughout Germany. Philip the Magnanimous had the letters carved on his furniture and emblazoned on the sleeves of his retainers. Many

Introduction

supporters of the Reformation had the letters engraved over their doorways in defiance of the Emperor and the Pope. But what are the letters doing in an English book on arithmetic? Was the author, Robert Recorde, or his printer Reynold Wolfe, or both of them in collusion, using VDMIE to covertly proclaim allegiance to the Protestant faith, in dangerous times when it was not politic to publicly declare oneself as either Protestant or Catholic? The woodcut was used again on the title page of the 1558 edition. By this time both Henry VIII and his successor, the protestant child king Edward VI, were dead. Henry's daughter, the staunch catholic Mary was the reigning monarch and during Mary's reign it is said that more than 300 protestant heretics were burnt at the stake. Perhaps it is no surprise then, that VDMIE has disappeared from the woodcut in this later edition.

THE PREFACE

Robert Recorde uses the preface to dedicate *The Grounde of Artes* to 'The ryght worshypfull mayster Rychard Whalley, Esquyer'. Whalley was in government service and when Henry VIII dissolved the monasteries, he visited the lesser houses on behalf of the king. In consequence he was granted church lands in 1539, and around this time he also purchased the Welbeck Abbey estates in Nottinghamshire. Later he was joint steward, under Protector Somerset during the minority of Edward VI, and crown receiver for Yorkshire. His fortune varied in later years and he was several times sent to the Tower on suspicion of embezzlement and treason. Nevertheless he surmounted his difficulties and became a very wealthy man before dying of old-age. Whalley was Recorde's patron and it is likely that at some time he was tutor to Whalley's children. This would not have been a sinecure, since Whalley had no less than twenty-five offspring – five by his first wife Laura, thirteen by his second wife Ursula and another seven

Introduction

by his last spouse Barbara. It is not unlikely that Recorde gained some of his earliest teaching experiences while in Whalley's household, and the scholar who features so prominently in his books may have been modelled on a particular favourite among Whalley's children. It may have been Whalley who was responsible for Recorde's introduction to court and to government service. Readers wishing to know more about Richard Whalley are referred to the *Oxford Dictionary of National Biography*.

NOTABLE PAGES

Perhaps the most notable pages of general interest are those in the section headed 'Reduction', where Recorde gives a table of English coins and then explains their values. For example, a Royal is worth one and half Angels, an old Noble, called a Henry, is worth 2 crowns, while a Noble, called a George, is worth 6s 8d. A Grote contains 4 pennies, but there is another Grote called a Harp which is only worth 3 pennies, and so on for many pages. Then comes weights and measures, where you will need your wits about you to decipher the word which is spelt 'haberdyepoyse'. Wool weights such as a todde and a sacke, cheese weights such as a clove and a wey, measures of liquor such as a pottell, a fyrken, a kilderkin, wine measures such as a rondelet, a hogs head and a pype are all explained together with the measures for herrings, salmon and eels. Then come dry measures like a peck, a bushel, a stryke, a cornoke, and the fact that 3 grains of barley make an inch, 3 feet 9 inches make an elle and five yards and a half make a perch. There is much of interest in these pages for even the most casual browser.

SOURCES

Readers wanting to know more about Robert Recorde and his famous series of mathematical books should consult the following:

Introduction

For an easily accessible biography visit the MacTutor History of Mathematics, 'Robert Recorde', [online]

http://www-history.mcs.st-andrews.ac.uk/Biographies/Recorde.html.

Written sources are:

Stephen Johnston, 'Recorde, Robert (c1512–1558)' *Oxford Dictionary of National Biography*, Oxford University Press, 2004.

Howell Lloyd, 'Famous in the Field of Number and Measure: Robert Record, Renaissance Mathematician', *Welsh History Review*, Vol. 2 (2000), pp. 254-282.

William Barr, 'A World View of Robert Recorde: A Brief Study of Tudor Cosmology, *Albion: A Quarterly Journal Concerned with British Studies*, Vol. 1, No. 1 (1969), pp. 1-9.

Joy B. Easton, 'The Early Editions of Robert Recorde's Ground of Artes', *Isis*, Vol. 58, No. 1 (Winter 1967), pp. 515-532.

Joy B. Easton, 'On the date of Robert Recorde's birth', *Isis*, Vol. 57, No. 1 (Spring 1966), p. 121.

Margaret E. Baron, 'A Note on Robert Recorde and the Dienes Blocks', *The Mathematical Gazette*, Vol. 50, No 374 (Dec 1966), pp. 363-369.

Louise Diehl Patterson, 'Recorde's Cosmography, 1556', *Isis*, Vol. 42, No. 3 (Oct 1951), pp. 208-218.

E.R. Sleight, 'Early English Arithmetics', *National Mathematics Magazine*, Vol. 16, No. 4 (Jan 1942), pp. 198-215 and Vol. 16 No. 5 (Feb 1942), pp. 243-251.

Francis R. Johnson & Stanford V. Larkey, 'Robert Recorde's Mathematical Teaching and the Anti-Aristotelian Movement', *The Huntingdon Library Bulletin*, No. 7 (Apr 1935), pp. 59-87.

David Eugene Smith & Frances Marguerite Clarke, 'New Light on Robert Recorde', *Isis*, Vol. 8, No. 1 (Feb 1926), pp. 50-70.

David Eugene Smith, 'New Information Respecting Robert Recorde', *The American Mathematical Monthly*, Vol. 28, No. 8/9 (Aug–Sep 1921), pp. 296-300.

Frank V. Morley, 'Finis Coronat Opus', *The Scientific Monthly*, Vol. 10, No. 3 (Mar 1920), pp. 306-308.

HERE BEGINS
THE GROUNDE OF ARTES

The groūd of artes

teachyng the worke and pra=
ctise of Arithmetike, moch necessary
for all states of men. After a more
easyer & exacter sorte, then any
lyke hath hytherto ben set
forth : with dyuers newe
additions, as by the
table doth partly
appeare.

ROBERT RECORDE.

The bokys uerdict.
To please, or dysplease, sure I am,
But not of one forte, to euery man.
To please the beste forte wold I say=
                                    (ne,
The frowarde to dysplease I am cer=
                                    (tayne.

The Table.

The contentys of the fyrſt Dialoge.

☞ The declaration of the profyte of Arithmetyke.

Numeration with an eaſy and large table.

Addition. ⎫
Subſtraction. ⎬ ẃ diuers exāples, and ſome new four mes of workynge.
Multiplicatiō, ⎪
Diuiſion. ⎭

Reduction with diuers declarations of coynes, weyghtes, and meaſures of ſondry fourmes, now newly ad= ded to theſe other rules.

Progreſſion both Arithmeticall and Geometricall, with certayne que= ſtions touchynge the ſame.

The golden rule, and the backer rule with diuers queſtions therto be= longynge.

The double rule of proportion.

The rule of felowſhyppe, bothe with tyme, and without tyme.

Unto all theſe is added theyr profe.

The Table

In the seconde Dialoge.

⁋ The fyrst .v. kyndes of Arithm=
tike wrought by counters.

The commen kyndes of castynge ac=
comptes, after marchautes fasshiō,
and auditours also.

Nombrynge by the hande, newly ad=
ded.

Finis.

The preface.

⁋ To the ryght worshypfull May=
ster Rychard Whalley Esquyer,
Robert Recorde wyssheth
health & prosperous
sucesse.

Ore oftētymes haue
I lamented with my
selfe the infortunate
conditiō of Englond,
seynge so many great
clarkes to aryse in sū=
dry other partes of the worlde, and so
fewe to appeare in this our nation :
where as for excellencye of naturall
wytte (I thynke) fewe nations do
matche Englysh men : But I can not
impute the cause to any other thyng,
then to the contempte or mysregarde
of learning. For as Englysh men are
inferyor to no men in mother wytte,
so they pass all men in vayne plea=
sures, to which they may attayne w=
out greate payne or labour : & are as
slacke to any neuer so great cōmodite
yf there hange of it any paynefull stu=
dye

The preface.

dye : how be it, yet all men are not of that sorte, though the moste parte be, the more pytie it is : but of them that are so glad not onely wͨ paynfull stu=dye & studyouse payne to attayne learnyng, but also wͨ as great studye and payne to cōmunicate theyr learnynge to other, & make all Englande (yf it mought be) partakers of the same, yͭ most part are such, yͭ vnneth they can supporte theyr owne necessary charges so that they are not able to beare any charges in doynge of that good, that els they desyre to do. But a greater cause of lamētatiō is this, that when learned men haue taken paynes to do thinges for the ayde of the vnlerned, scarse they shall be alowed for theyr well doyinge, but deryded and scorned and so vtterly dyscouraged to take in hand any lyke enterpryse agayne. So yͭ yf any be found (as there are some) that doth fauour learnyng & learned wyttes, and can be content to forther learnyng, yea onely with theyr word,
suche

The preface.

such mē, though they be rare, yet shal they encourage learned men to enterpryse some thynges, at the least, that Englond may reioyse of. And I haue good hope ẙ Englond wyll (after she hath taken some sure taste of lerning) not only bryng forth more fauourers of it, but also such learned men, ẙ she shall be able to cōpare ẘ any realme in the world. But in ẙ meane ceason, where so few regarders of lernynge are, how greatly they are to be estemed that doth fauour and forther it, my penne wyll not suffyce to declare. Therfore good M. Whalley, where as I do vpon iust occasyō iudge, yea & know you to be one of them ẙ both loued & also moch despzeth to further good learnyng, and yet am not well able to write your cōdygne prayse for the same, I thynke it better with sylence to ouerpasse it, then other to say to lytle of it, or to prouoke agaynste you the malyce of them, which do nothynge them selfe ẙ is prayse worthy,

and

The preface.

and therfore can not abyde to heare the prayse of any other mannes good dede, & consyderyng your great fauour vnto learnyng, though I my selfe be not worthy to be reckened in the nomber of greate learned men, yet am I bolde to put my selfe in presse wt such abylyte as God hath lent me, though not with so greate conynge as many men, yet with as greate affection as any man to helpe my countre men, & wyll not cease dayly (as much as my small abylyte wyl suffre me) to endyte some suche thynge that shal be to the enstructiō, though not of learned mē, yet at the leaste of the vulgare sorte: whose argument all wayes shall be such, that it shall delyte all learned wyttes, thoughe they do not learne any great thynges out of it. But to speake of this present boke of Arithmetike, I dare not nor wyll not set it forth with any wordes, but remytte it to the iudgement of all gentyll readers, and namely to you (good M. Whalley

The preface.

Whalley) besechyng fyrste you so to esteme it, as it doth seme worthy. And so, other to accept ẏ thyng for it selfe, other at the leaste to allowe my good endeuoure: but I pceaue I nede not vse any persuasyons vnto you, whose gentell nature and fauourable mynd is redye to receyue thankefully, and enterprete to the beste all such enterprises attempted for so good an ende, though the thynge do not all wayes satysfye mēnes expectation. This cōsydered dyd bolden me to approche vnto you with this lyttell boke of the arte of numbrynge, which yf you shal receyue fauorably, you shal encorage me to visite you hereafter with some greater thynge. And as I iudge you of so louyng a mynde to your natyue countrye, that you wold moch reioyse to se it to prospere in good learnynge and wittie artes, so I hope well of al Englysshe men, that they wyll be not vnmyndfull of your dewe prayse, by whose meanes they are helped & furthered

The preface.

thered in any thynge : nother oughte they to esteme this thynge of so lytell valewe, as many men of lyttell dyscretion oftentymes do : for who so setteth small pryse by the wyttie deuyse and knowledge of numbryng, he lyttell consydereth it to be the chyefe poynt (in maner) wherby men dyffer from all brute beastes : for as in all other thynges (all most) beastes are partakers with vs, so in numbryng they dyffer clene from vs.

The foxe in crafty wyt excedeth most men.
A dogge in smellyng hath no mā his pere.
To foresyghte of wether yf you loke then,
Many beastes excelle man this is clere.
ẏ wittines of elephāts doth letters attayne
But what cōnyng doth there in the bee remayne?
The emmette forseyng ẏ hardnes of witer.
Prouideth vitayles in tyme of sommer.
The nyghtyngale, the lynet the thrusshe, the larke,
In musical harmony passe many a clarke.
The hedgehogge of astronomy semeth to knowe. And

The preface.

And stoppeth his caue where the winde wyll blowe. Many thynges els of wittynes of beastes & byrdes myght I here saye (saue that an other tyme of them I entede to write) wherin they excell in maner all men, as it is dayly sene: but in nōber was there neuer beaste founde so cōnyng, y̆ coulde know or discerne one thynge from many, as by dayly experyence you maye well cōsyder, when a bytche hath many whelpes, or a henne many chyken, & lyke wayes of other, what so euer they be, take frō them all theyr yonge, sauyng onely one, & you shall perceyue playnly y̆ they mysse none, though they wyll resyste you in takynge them away, and wyl seke them agayne, yf they maye knowe where they be: but els they wyll neuer mysse them truely, but take away that one that is lefte, and then wyll they crye and complayne: and restore to them y̆ one, then are they pleased agayne, so that of nomber this I may iustly say,
it is

The preface.

it is thonely thyng (all most) that se=
parateth man from beastes: he ther=
fore that shall contemne nōber, he de=
clareth hym selfe as brutysshe as a
beaste, and vnworthy to be counted
in the felowshyp of men. But I truste
there is no man so fowle ouersene,
though many ryghte smally do it re=
garde. Therfore wyll I now staye to
wryte agaynst such, & returne agayne
to this my boke, whiche I haue wryt=
ten in ẙ fourme of a dyaloge, bycause
I iudge that to be the easyest waye of
enstructiō, when the scholer may aske
euery doubte orderly, and ẙ mayster
may answere to his questiō playnly.
Now be it, I thynke not the contrary,
but as it is easyer to blame an other
mans worke, then to make the lyke,
so there wyll be some that wyll finde
faute, bycause I wryte in a dyaloge:
but as I coniecture, those shall be
suche as do not, can not, other wyll
not perceaue the reason of ryght tea=
chynge, and therfore are vnmete to
be

The preface.

be farther answered vnto, for suche men with no reason wyll be satiſfyed. And yf any man obiecte that other bokes haue ben wrytten of Arithme= tyke all redy ſo ſufficiently, that I ne= ded not now to put penne to the boke, except I wyll condempne other men= nes wrytynges : to them I aunſwere, that as I condempne no mans dyly= gency, ſo I knowe that no one man can ſatiſfye euery man : and therfore lyke as many doth eſteme greatly o= ther bokes, ſo I doubte not, but ſome wyll lyke this my boke aboue any o= ther Englyſhe Arithmetike hetherto wrytten, & namely ſuche as ſhall lacke inſtructers, for whoſe ſake I haue ſo playnly ſet forthe the examples, as no boke (that I haue ſene) hath done hetherto, which thyng ſhall be great eaſe to ẏ rude reader. Therfore good M. Whalley, though this boke can be vnto your ſelfe but ſmall ayde, yet ſhall it be ſome helpe vnto your yong chyldren, whoſe fortheraunce you de=
ſyre

The preface.

syze no lesse then your owne. And though to you pryuately I do it dedicate, yet I doubte not (such is your gentelnesse) but that you can be contente that all men vse it, and employ the same to theyr moste profyte: which thyng yf I perceaue that they thankfully do, and receaue it with as good wylle as it was wzitten, then wyll I shortly with no lesse kyndnes set forth such entroductious into Geometry & Cosmography, as hytherto in Englyshe hath not ben enterpzysed, wher with (I dare saye) all honeste hartes wil be pleased, and all studyouse wyttes greately delyted. I wyll saye no moze, but let euery man iudge as he shall se cause. And thus for this tyme wyll I staye my penne, commyttynge both you & all yours (good M. Whaleye) to that true fountayne of perfect nōber, which wrought the hole world by nōber, & measur: he is trinite in vnite, & vnite in trinite: to whome be all prayse, honour & glozye, Amen.

¶ Before the initoduction of Arithmetike these fi=
gures muste be learned.

¶ Figures of nomber.

i.	one	xx.	twentye.
ii.	two	&c.	
iii.	thre	xl.	fourthy.
iiii.	foure	l.	fyftye.
v.	fyue.	lx.	syxtye.
vi.	syxe.	&c.	
vii.	seuen.	xc.	nyntye.
viii.	eyghte	C.	a hundred.
ix.	nyne.	CC.	two hundred.
x.	ten.	&c.	
xi.	aleuen.	D.	fyue hundred.
&c.		DC.	syxe hundred.
xv.	fyuetene.	&c.	
&c.		M.	thousande.
xix.	nynetene.		

☞ Figures of monye.

☞ c. a cee, the .xvi. ⎫
q.   a kewe, the .viii. ⎬ part of .i. pēny
q̈   a farthing, p̊ .iiii. ⎭
ob. an halfe pennye.
i. d̀. a pennye.
i. ṡ. a ſhyllynge.
i. li. a pounde.

¶ A dyaloge betwene the Mayster
and the Scoler : teachynge the
arte and vse of Arithmetike
with the penne.
   Scoler.

Yr such is your
auctorite in myne
estimatiō, that I
am content to con
sente to your say=
enge, and to re=
ceaue it as truth :
though I se none
other reason, that doth leade me ther=
vnto : where as els in myne owne cō=
ceyte it appereth but vayne, to be=
stowe any tyme priuately in lernyng
of that thyng, that euery childe may
and doth learne at all tymes & hours
when he doth any thyng hym selfe a=
lone, & moch more when he talketh
or reasoneth with other.   Mayster.
   Lo, this is the fashyon and chaūce
of all them, that seke to defende theyr
blynde ignoraunce. That when they
            A   thinke

The commodities

thynke they haue made ſtronge rea=
ſon for them ſelfe, then haue they pro=
uyd the, quyte contrarie. For if nūbe=
rynge be ſo cōmen as thou grauntest
it to be, that no man can do any thing
alone, & moch leſſe talke or bargayne
with other, but he ſhall ſtyll haue to
do with numbre: this proueth not nū=
bre to be contemptyble & vyle, but ra=
ther ryght excellent, and of hygh re=
putation ſyth it is the ground of all
mēnes affayres, ſo that without it no
tale can be tolde, no communication
without it can be longe cōtynued, no
bargaynyng without it can duely be
endyd, nor no buſynes that man hath
iuſtly completed. Theſe cōmodities,
if there were none other, are ſufficiēt
to approue the worthines of nombre.
But there are other vnnumerable
farre paſſyng all theſe, which declare
nombre to exceade all prayſe. Wher=
fore in all greate workes, are clarkes
ſo moch deſpyzd? wherfore are audy=
tours ſo rychely feyd? What cau=
ſeth

of Arythmetyke.

seth geometrians so hyghly enhaun=
cyd? why are astronomers so greatly
aduauncyd? bycause that by numbre
suche thynges they do kynde, whiche
elles shuld farre excelle mans mynde.
Sco. Merely syr if it be so, that these
men by nomberynge theyr conyng do
attayne, at whose great workes most
men do wonder, then I se well I was
moch deceaued, and nomberynge is a
more connyng thynge than I toke it
to be. M. Yf nombre were so vyle
a thynge as thou dyddest esteme it,
then nede it not to be vsed so moch in
mens comunycation. Exclude nom=
bre and answere me to this question:
Howe many yeares olde arte thou?
S. Mum. M. Howe many dayes in a
weke? how many wekes in a yere?
what landes hath thy father? howe
many men doth he kepe? how longe
is it syth you came from hym to me?
Sco Mum. M. So that yf nombre
wante, you answere all by mumes:
Howe many myle to London? Sco.
              A.ii.   A poke

The commodities

A poke full of plumbes. Ma. Why thus maye you se what rule nombre beareth, and that yf nombre be lackynge, it maketh men dumme, so that to most questions, they must answere mum. S. This is the cause syr, that I iudged it so vyle / bycause it is so comen in talkynge euery whyle : For plenty is no denty, as the comen sayenge is. M. No, nor store is no sore : perceaue you this? The more comen that a thyng is, beynge nedefully requyred, the better is the thynge, and the more desyred. But in nomberyng as some of it is lyght and playne, so the most parte is dyfficulte, and not easye to attayne. The easyer parte serveth all men in compn, and the other parte requyreth some lernyng. Wherfore as without nomberynge a man can do almost nothynge, so with the helpe of it, you maye attayne to all thyng. S. Yea syr? why then it were beste to learne the arte of nombrynge fyrste of all other lernynge, and then
a man

of Arythmetyke.

a man nede learne no more, yf all other come with it. M. Nay not so, but yf it be fyrst learned, then shall a man be able (I meane) to lerne, perceaue, and attayne to other sciences, which without it, he shulde neuer gette. S. I perceaue by your former wordes, that Astronomye and Geometrye depend moch of ỹ helpe of nombrynge. But that other sciences, as musyke, physyke, lawe and grammer, & suche lyke, haue any helpe of Arythmetike I perceaue not. M. I may perceaue your great clarkelynes, by the orderynge of your sciences, but I wyll let that passe nowe, bycause it toucheth not the matter that I entende, and I wyl shewe you how Arithmetike doth profitte in all these, somwhat grosly, acccordynge to your small vnderstandynge, omyttyng other reasons more substancyall. Fyrste (as you rekened them) Musyke hath not only greate helpe of Arithmetyke, but is made, & hath his perfectnes of it. For all mu-

A.iii.   syke

The commodities

Arithmetyke necessary for Physike.

syke standeth by nombre and propor=
tion. And in physike, besyde the calcu
lation of criticall dayes, with other
thynges, whiche I omytte. How can
any man iudge the pulse rightly, that
is ignoraunt of the proportion of nō=
bres?

The Lawe.

And as for the lawe, it is playne
that ye man, that is ignorant of Arith
metyke, is nother mete to be a iudge,
nother a proctour. For howe can he
well vnderstand another mans cause
apperteynyng to distribution of goo=
des, or other dettes or sūmes of monie
yf he be ignoraunt of Arythmetyke?
This oftentymes causeth right to be
hyndered, when the iudge other de=
liteth not to heare of a matter that he
perceaueth not, other can not iudge
it, for lacke of vnderstandynge, and
this cōmeth by the ignoraūce of arith

Grammer.

metike. Now as for grāmer, me thin=
keth thou sholdest not dowte in what
it nedeth nōbre, syth thou hast lerned
that nownes of al sortes, pronownes,
verbes, and participles are distincte
dyuersly

of Arythmetyke.

dyuersly by nombers: besydes the va
riete of nownes of numbre, & aduer=
bes. And if you take awaye nombre
from grammer, then is all the quan=
tite of syllables loste. And many
other wayes doth nomber helpe
grāmer. Wherby were all kyndes of
meter found and made? Was it not
by nomber? But how nedefull arith=
metike is to all partes of philosophie   *Phylosophie.*
they maye sone se, that readeth other
Aristotle, Plato, or any other phylo=
sophers wrytinge. For all theyr exam
ples (all most) and theyr probations
depende of Arithmetike. It is the
sayenge of Aristotle, that he that is
ignorant of Arithmetike, is mete for
no scyence. And Plato his mayster
wrote a lyke sentence ouer his scole=
house doore: Let none enter in hyther
(φhe) that is ignorant of geometrie,
Seyng he wolde haue all his scolers
experte in geometrie, moch rather he
wolde the same in Arithmetike, with=
out whiche, geometrie can not stand.

                A.iiii.   And

The commodities

And how nedefull Arithmetike is to diuynitie, it appereth, seyng so many doctors gatheryng so great mysterys out of numbre, and so moche to write of it. And if I shuld go about to write all the cōmodities of Arithmetike in cyuyll actes, as in gouernaunce of cōmyn weales in tyme of peace: and in dewe prouision and order of armes in tyme of warre. For nombryng of thoste: sūmynge of theyr wages: prouysions of vitayles: beiwyng of artillarie, with other armour. Besyde the connyngest poynte of all: for castyng of ground: for encāpyng of men, with other lyke. And howe many wayes also Arithmetike is conducyble for al priuate weales, of lordes and all possessioners, of marchauntes and all other occupyers: and generally, for all estates of men besides auditours, treasourers, receyuers, stewardes, baylyffes, & such lyke, whose offices without Arithmetike is nothyng: yf I shulde (I saye) particularly repete all

of Arythmetyke.

all such cōmodities of this noble sci=
ence of Arithmetike, it were ynough
to make a very greate booke. S. No,
no, syr you shal not nede : for I dowte
not, for this that you haue sayd, were
enowgth to perswade any manne, to
thynke this arte to be right excellent
and good and so necessarie for man,
that (as I thynke nowe) so moch as
a man lacketh of it, so moche he lac=
keth of his sense & wytte. M. What,
are you so farre chaunged syns, by
hearynge the fewe cōmodities in ge=
nerall? by lykelyhode you wolde be
farre chaunged yf you knewe all the
cōmodities particular. S. I beseche
you syr reserue those commodities,
that reste yet behynde, vnto theyr
place more cōuenient. And if you wyl
be so good as to vtter at this tyme
this excellent treasure, so that I may
be somwhat enryched therby, and yf
euer I shall be able, I wyll requyte
your payne. M. I am very glad of
your requeste, and I wyll do it spe=
A.v.   dely

The commodities

dely, sith that to learne it, you be so reddye. S. And I to your auctorite, my wyttes do subdewe: what so euer you say, I take it for trewe. M. That is to moch, and mete for no man, to be beleued in al thynges, without shew=ynge of reason. Though I myght of my Scoler some credence requyre, yet except I shewe reason, I do not it de=syre. But now syth you are so earnest=ly sette this arte to attayne, best it is to omitte no tyme, lest some other pas=syon coole this great heate, and then you leue of before you se the ende. S. Though many there be so vnconstāt of mynde, that flytter and turne with euery winde, which often begyn and neuer come to the ende, I am none of theyr sorte, as I truste you partely knowe. For by my good wyll what I ones begin, tyll I haue it fully ended I neuer blynne. M. So haue I foūd you hetherto in dede, and I trust you wyll encrease rather then go backe, for better it were neuer to assay, then

to

of Arythmetyke.

to shrinke & flee in the myddell waye, but (I truste) you wyll not so do / therfore tell me brefely, what call you the science, that you despze so greatly. S. Why sir, you knowe. M. That maketh no matter, I wolde heare, whether you knowe and therfore I aske you. For greate rebuke it were, a science to haue studied, and yet can not tell how it is named.

S. Some call it Arsmetrike, and some Awgryme. M. And what dothe those names betoken? S. That (yf it please you) of you wolde I learne.
M. Bothe names are corruptly writen, Arsmetrike for Arithmetyke (as the Grekes call it) and Awgrym for Algorisme (as Arabyans sounde it) whiche bothe betoken the science of nombrynge. For arithmos in greke is called nomber, & of it cometh arithmetike, the arte of nomberynge. So that arithmetike is a science or arte, teachynge the maner and vse of nombrynge, & maye be wrought dyuersly
with

Arithmetyke.

Ἀριθμηπκή.

Ἀρωμός.

The commodities

with penne or couters, & other wayes But I wyll fyrst shewe the workynge with the penne, and then thother in order. *Sco.* This I wyll remembre. But how many things are to be learned, to attayne this arte fully? *M.* There are rekened comenly .vii.

☙ Numeration, Addition, Subtraction, Multiplication, Diuisiō, Progression, and Extraction of radicals: to these, some men adde Duplacyon, Triplacyon, and Mediatiō. But as for these last thre, they are conteyned vnder the other .vii. For Duplacion and Triplacion, are conteyned vnder Multiplication, as it shall appere in theyr place. And Mediation is conteyned vnder Diuision, as I wyll declare in his place also. *S.* Yet then there remayne the fyrste .vii. kyndes of nombrynge. *M.* So there dothe: Howbeit yf I shall speake exactely of partes of nombrynge, I must make but .v. of them: For Progression is a compounde operation of Addytion, Multy=

of Arythmetyke.

Multiplication, and Dyuision. And so is the Extraction of Radicals.

But it is no harme to name them as kyndes seuerall, seynge they appere to haue some seuerall workynge. For it forsethe not so moche to contende the nomber of them, as for the dewe knowledge and practisynge of them. Sco. Then you wyll that I shall name them as .vii. kyndes distincte. But nowe I desyre you to enstructe me in the vse of eche of them. Ma. So wyll I, but it muste be done in order: for you may not learne the last as soone as the fyrste, but you muste learne them in that order as I dyd reherse them, yf you wyll learne them spedely and well. S. Euen as you please. Then to begyn, Numeration is the fyrste in order, what shall I do with it. M. Fyrste you muste knowe what the thinge is, and then after learne the vse of the same.

Nume=

Numeratyon.

Umeratiō is the arte to expresse and rede all sommes proposed, and is of two sortes, for other it gethereth the valewe of a some proposed, other els it expresseth a sum conceaued by figures & places dewe. S. Why? then me thynketh you put a dyfference betwene the valewe and the figures. M. Yea, so do I. For the valewe is one thyng, and the figures are an other thynge, and that cōmeth partely by the dyuersite of fygures, but chefely of the places, wherbe thei be sette. S. Then I must knowe here iii. thynges, The valewe, the fygure and the place. M. Euen so : but yet adde order to them, as the fourthe. And fyrst marke that there are but .x. figures, that are vsed in arithmetike, and of those .x. one doth sygnifie no=thyng, which is made lyke an o, and is called priuately a cyphar, though all the other somtyme be lykewyse named

A Cyphar

Numeration.

named. The other .ix. are called syg‸ nifienge figures, ⁊ be thus figured: *Figures.*

  1 2 3 4 5 6 7 8 9

And this is theyr valewe.

i. ii. iii. iiii. v. vi. vii. viii. ix.

⁋ But here muste you marke, that euery fygure hath .ii. valewes. One alwayes certayne, that it sygnifieth properly, which it hath of his forme. And the other vncertayne, whiche he taketh of his place.

⁋ A place is called the seat or roume *a Place.* that a fygure standeth in. And loke howe many fygures are wrytten in one summe, so many places hath that hole valewe. And the fyrst place must be called p̃, that is next to the righte hande, and so reckenyng by order to‸ warde the lefte hande, so p̃ that place is least, that is nexte to the left hand. As for example, yf there stode before you syxe men in a rowe, syde by syde ⁊ you shulde tell them as they stande in order, begynnynge with the man that were next to your ryght hande,

than

Numeratyon.

than he that were nexte hym, shuld be called the seconde, and so forth to the fartheft frō your ryght hande, which is the .vi. & the laste. S. Syr I perceaue you well, so myght I recken letters or any other thynge. As yf I sholde wryte .viii. letters after this order, a, b, c, d, e, f, g, h, nowe muſte I saye, h. is the fyrſt, g. the .ii. f. the iii. e. the .iiii. .d. the .v. c. the .vi. b. ẏ vii. and a. the .viii. M. That is well done. And after ẏ same sorte vse here after, so that what I declare by one example, do you expresse by an other, and so shall I perceaue whether you vnderſtand it or no. And so paſſe ouer nothynge tyll you perceaue it well, & be experte therin. S. Syr, I praye you, howe many of these places be there in all. M. There is no certayne nūber of them, but they are somtimes more, and somtymes fewer, accordyng to the summe that is expreſſed. For so many as the figures are, so many are the places, and the laſt place is so cal
led

Numeration.

led, not bycause it is the laste of all other, but is the laste of that present summe / and it maye be the myddell place in an other summe. S. Me semeth I perceaue this very well, as towchynge thorder of reckenynge of the places. But as for the nomber of them, you say there is no certayntie. Now there resteth to declare the valewe of the fygures, by dyuersitie of places, which you called the valewe vncertayn. M. But fyrst let me heare whether you knowe perfectly the certayne valewe. S. Yes syr, as you wrote them, so I marked them. M. How write you then .v? S. By this figure 5. M. And how .vi? S. Thus 6. M. Write these .iii. nombers eche by it selfe, as I speake them .vii. .iiii. .iii. S. 7. 4. 3. M. How write you these foure other .ii. .i. .ix. .viii? S. Thus (I trowe) 2. 1. 6. 8. M. Nay, there you mysse : Loke on myne exãple agayne. S. Syr trouth it is, I was to blame, I toke 6 for 9, but I wyll be warer

Valewe un certayn.

B   here

Numeratyon.

here after. M. Now then take hede, these certayne valewes euery fygure repzesenteth, when it is alone wzitten without other figures ioyned to him And also when it is in the first place, though many other do folowe, as foz example, This figure 9 is .ix. stan= dyng now alone. S. Now is he alone and standeth in the myddell of so ma= ny letters? M. The letters are none of his felowes. And yf you were in Fraūce in the myddle of M. Frenshe men, yf there were no Englyshe man with you, you wold recken your selfe to be alone. S. So it is. Then 9 with= out moze figures of Awgrym, beto= keneth .ix. what so euer other letters be about it. M. Euyn so, and so doth it, yf it be in the fyzst place ioyned w other, how many so euer do folowe, as in this example 3679, you se 9 in the fyzste place, and doth betoken .ix. as yf he were alone. S. J perceaue ẏ. And doth not 7 stande in the second place, and betoken. vii? and 6 in the
thyzd

Numeratyon.

thyrde place, and betoken. vi? And so 3 in the fourth place, & betokyn. iii.
M. Theyr places be as you haue said but theyr valewes are not so. For as in the fyrste place, euery fygure betokeneth his owne valewe certayne only. So in the second place euery figure betokeneth his owne valewe certayne .x. tymes: as in that example, 7 in the seconde place, is .vii. tymes .x. that is .lxx. And in the thyrde place, euery figure betokeneth his owne valewe a hundreth tymes, so ẏ 6 in that place, betokeneth .vi. hundred. And in the .iiii. place, euery figure betokeneth his owne valewe, a M. tymes as in the foresayd nōber: 3 in the .iiii. place standeth for .iii.M. And in the .v. place, euery figure standeth for his owne valewe .x.M. tymes. And in the .vi. place a C.M. tymes. And in the .vii. place, a M.M. tymes. And in the .viii. place .x.M. M. tymes: so that euery place exceadeth the former .x. tymes. S. As thus

B.ii. yf

Numeratyon.

yf I make this nombre at all aduentures, 91359684 here are .viii. places, In ẏ fyrſt place is 4, and betokeneth but .iiii. In the ſecond place is 8, and betokeneth .x. tymes .viii. ẏ is .lxxx. In the thyrde place is 6, and betokeneth .vi.C. In the fourth place 9. is .ix.M. And 5 in the .v. place, is .x. M. tymes 5, that is .l.M. So 3 in the .vi. place, is a C.M. tymes 3, ẏ is CCC.M. Then 1 in the .vii. place a MM. And 9 in the .viii. place .x. M.M. tymes 9, that is .xc.M.M. But now I can not eaſely, nor quyckely rede it in order. M. That ſhall you practyſe by this meanes. Fyrſte put a prycke ouer the fourth figure, and ſo ouer the .vii. And yf you had ſo many, ouer the .x. xiii. xvi. and ſo forth, ſtyll leuing .ii. figures betwene eche .ii. pryckes. And thoſe roumes betwene the pryckes, are called, ternaries. Then begyn at the laſt prick, & ſe howe many figures are betwene hym, & the ende, whiche can not paſſe iii.

Ternarye, or Trinite.

Numeration.

iii, rekenynge hym selfe for one, then pronounce them as yf they were writ ten alone from the reste, and adde at the ende of theyr valewe, so many ty= mes thousande, as your nombre hath pryckes. Then come to the nexte .iii. figures, and sounde them as yf they were aparte from the reste, and adde to theyr valewe so many tymes thou= sandes, as there are prickes betwene them and the fyrst place of your hole nombre. And so do by euery other .iii. figures folowynge, yf you haue mo, as in exāple 91359684. this was your nombre : put a prycke ouer 9 in the iiii. place, & ouer 1 in the .vii. place, & then no more (for your places come not to tenne) as thus 91359684. Now go to the last pricke ouer 1, and take it, and the figure 9 that foloweth it, and valewe them alone. S. 91 that is xci. M. So is it, but then adde for the nomber of your prickes, twyse M. S. That is .xci. thousand thousand. M. So is it. Now take the .iii. other

B.iii. figu=

Numeration.

figures from 1 to ye next pricke, & va=
lewe them. S. 359 that is .CCC.lix.
M. Now adde for the one pricke, that
is betwene them & the fyrst place .M.
S. CCC.lix. thousand. M. Now come
to the other .iii. figures that remayn.
S. 684, that is .vi.C.lxxxiiii. M. Now
haue you valewed all. And at ye ende
of this laste nomber, you shall adde
nothynge, bycause there remayneth
no prycke nor nomber after it : yet
proue in an other nomber, as thus,

$$230864089105340.$$

Scoler. 230864089105340. I haue
prycked them, as you taught me / but
I am in doubt whether I haue done
well or no, bycause of the cyphars : for
I remembre, you tolde me that they
do signifie nothynge, and therfore I
doubte whether I shuld recken them
for a figure, in settyng of the prickes,
and agayne I knowe not wherfore
they serue. M. That wyll I tell you
now : in dede they are of no valewe
them selfe, but they serue to make vp
nomber

Numeration.

nombre of places, and so maketh the figure folowynge them to be in a forther place, and therfore to signifie the more valewe, as in this example, 90 the cyphar is of no valewe, but yet he occupieth the fyrst place, and causeth 9 to be in the seconde place, and so to signyfie .x. tymes 9, that is .xc. so ẏ ii. cyphars thrusteth the fygure folowynge them, in to the .iii. place, & so forth. S. Then I perceaue in the example aboue, I haue prycked well ynough, for though that cyphar, that is pricked, signifie nothyng, yet must he haue the prycke, bycause he came in the .xiii. place. Then wyll I proue to nombre that summe : fyrst there is 230. M.M.M. 864 M.M.M. And what shall I now do? there is a chyphar in ẏ .iii. place, and no figure after hym, but they that I haue reckened. M. He dyd serue for them, that you haue all redy reckened, to make thē in a place forther, then they shuld be, yf he were awaye, & therfore nowe

B.iiii.   you

Numeration.

you shall lette hym go. And so do al=
wayes when he occupyeth the place
nexte before any prycke, which is the
laste of that ternarie: and a cyphar in
the last place doth nothyng. S. Then
I shall say but 89 M.M. M. No, but
go forth. S. 105 M. Now are all my
pryckes spent, and yet remayne 340,
so I must valewe them. CCC.xl. on=
ly. M. Now can you recken after this
sorte. And remembre, that euery suche
roume so parted, is called a ternarie,
or trinite. Some do parte such great
nōbres with letters, after this maner
2 3 0 ͨ8 6 4 ͨ0 8 9 ͨ1 0 5 ͨ3 4 0. In whiche
exāple ye may se that a, supplyeth the
roume of your pricke. And some doth
parte the nombres with lynes after this
fourme, 230|864|089|105|340.
where you se as many lynes, as you
made pryckes, and to one entent, saue
that ỹ lynes doth more playnly parte
euery .iii. figures, accordyng as they
shulde be valwed vnder one denomi=
nation. S. Yea syr, but yf you shulde
shewe

Numeration.

shewe me a nōber so parted, I shulde take it for many nōbres, and not for one. M. So might you do, not knowynge my meanynge. But what yf I dyd set forth the nomber without lynes, and your selfe (for the ease of rekenynge) dyd so parte it with lynes, wold you forgette wherfore you dyd it, and then take them for many nombres? S. No, I trowe not, but yet I doubte. M. Then vse that, that you lyke beste, for all the thre wayes is to one entent, saue (as I sayde) that the lynes doth more playnly distincte the denominatiōs. S. What call you denominations? M. It is the laste valewe or name added to any sume. As when I say .CC.xxii. poundes : poundes is the denominatiō : & lyke wayes in sayenge 25 men, men, is the denomination, and so of other. But in this places, that I spake of before, ẏ laste nomber of euery ternary, is the denomination of it. As of the fyrst ternary the denomination is vnities, and of

B.v.   the

Numeration.

the seconde ternary, the denominatiō is thousandes: & of the thyrd ternary thousand thousandes, or myllyons: of the .iiii. thousand thousand thousādes or thosand myllyons, & so forth. S. And what shall I call the valewe of the .iii. figures that maye be pronoūced before the denominators? as in sayeng 203000000 that is CC.iii. myllions, I peceue (by your wordes) that myllions is the denomination, but what shal I call the .CC.iii. ioyned before the myllions? M. That is called the numeratour or valewer, & the hole summe that resulteth of thē bothe, is called the summe, valewe, or nomber. S. Now, is there any thyng els to be lerned in Numeratiō? or els haue I lerned it fully? M. I mought here shewe you, who were the fyrste inuentours of this arte, and the reasons of all these thinges that I haue taught you, as why you shuld reckē your order of places backewarde, I meane from the ryght syde towardes the

Numeratour.

Summe.

Numeration.

the lefte, with many other thynges, touchynge the causes and reasons of it, but that wyl I reserue tyll ye haue lerned ouer all ye practise of this arte, least I shuld trouble your witte with ouer many thynges at the fyrst. But yet this must you marke, ye there are iii. kyndes of nomber, one called dy= *Thre kyndes of nomer.* gettes, an other articles, and ye thyrd myxte numbre. A dyget is any nūber *Digete.* vnder 10, as this 1. 2. 3. 4. 5. 6. 7. 8. 9. And 10 with all other that may be de uided in to .x. partes iust (& nothing remayne) are called articles, such are *Article.* 10. 20. 30. 40. 50. &c. 100. 200. &c. 1000. &c. And that number is called myxte *Mixte.* that contepneth articles, or at ye least one article, and a digette, as 12. 16. 19. 21. 38. 107. 1005. and so forth. And for the more ease of vnderstandynge and remembraunce, marke this : The di= gette number is neuer written with more than one figure, but the article and the myxte number are euer writ= ten with more then one figure : & thus they

Numeration.

they differ, that the article hath euer=
more this ciphar o, in his fyrst place,
and the myxte nūber, hath euer there
some diget. S. By this laste wordes,
I perceaue it moch better then I dyd
before, & now (I thynke) I wyll ne=
uer mysse to know those .iii. a sonder.
M. Yf you remember now all that I
haue saide, you haue learned sufficy=
ently this fyrst kynde of Arithmetike
called Numeration. Howbeit I wyll
yet exhorte you now, to remēber both
this, that I haue said, and all that I
shall saye, & to exercyse your selfe in y͞e
practise of it : for rules without pra=
ctise, is but a lyght knowledge, and
practise it is, that maketh men prefecte
and prompte in all thynges. And as
you haue learned to gether the va=
lewe of a sum͞e proposed, so must you
practise to expresse any sum͞e w͞t theyr
dewe figures : as for a profe : How ex=
presse you this summe, fyue thousand
two hundred, fyfty & seuen? S. This
troubleth me now, whether I shulde
begyn

Numeration.

begyn at the fyrst figure, or at y̆ last / for reason (me thynketh) shuld cause me to begynne at the fyrst, and yet yf I writte it as you spake it, I must be gyn at the last. M. When you knowe your places pfectly, you maye begyn where you lyste : but the more ease for your hand, is to begyn with the last, that is to saye, as I dyd speke them. But for the more suretie : a whyle you may begyn with the fyrst, repetynge my wordes backwarde, thus : Seuen fyfty, two hundred, fyue thousande, Or els sowndyng them al by theyr di get or valower, as thus : Seuen, fiue, two, fiue, for that waye is easyst. But then must ye loke well, whether there be any cyphar in your summe, that he may be set in his place, as yf your last valower of your sum̄e (as you speke it) be aboue 9, then is there a cyphar in the fyrst place. And yf it be a hun= dred or aboue, thē is there .ii. ciphars, one in the first place, and an other in the second, and so forth. But bycause
this

Numeration.

this thyng is such that can not be set forth wout many wordes, I thynke beste here now at the ende of Nume̱ration, to adde a table easy and reddy for the first exercise of it. Lo this is that table.

i. M. of mil.	M. of mil.	C. of mylli.	i. of myllios	Myllions.	C. of thousā.	i. thousādes.	Thousandes.	Hondredes.	Tennes.	Unites.	The denominaͤtors of y̱ plaͤces, or valews vncertayne.	
9	9	9	9	9	9	9	9	9	9	9	nyne.	
8	8	8	8	8	8	8	8	8	8	8	eight.	The names of the dygettes, vaͤlewes certayne, or valowers.
7	7	7	7	7	7	7	7	7	7	7	seuen.	
6	6	6	6	6	6	6	6	6	6	6	syxe.	
5	5	5	5	5	5	5	5	5	5	5	fyue.	
4	4	4	4	4	4	4	4	4	4	4	foure.	
3	3	3	3	3	3	3	3	3	3	3	three.	
2	2	2	2	2	2	2	2	2	2	2	two.	
1	1	1	1	1	1	1	1	1	1	1	one.	
0	0	0	0	0	0	0	0	0	0	0	cyphar.	
Eleuenth.	Tenthe.	Nynthe.	Eyghteth	Seuenth.	Syxte.	Fyfte.	Fourthe.	Thyrde.	Seconde.	Fyrste.	The orͤder of y̱ places	

(The lefte syde or hande. / The ryght syde or hande.)

Numeration.

℃ This table (as yow maye se) hath aleuen places, & in eche of them are set all the digettes, whose certayne valewe is wrytten on the ryght hand of the table, and the valewe vncertayne on the lefte hande. So that by this table you maye lerne bothe how to expresse any number that you lyste (yf that it excede not .xi. places) that is to saye .lxxxx. thousand myllyons, and so may you by the helpe of it, valewe all summes proposed vnder the sayde number : as for exāple, the sūme that I proposed before, whiche was fyue thousand, two hundred, fyfty & seuen, yf you wyll expresse it, take the fyrste number (as I speke it) whiche is .v.M. whose valower or certayne valowe is .v. and his vncertayne valowe or denomination is .M. Fyrste you shall seke at the ryght hande, the valewer .v. Then seke alonge vnder the tytle of denomination, towarde the lefte hande, tyll ye fynde thousades, and vnder it ryght at the fote of the

Numeration.

the table, is the number of the place, that is the fourth, wherein you muste write your dyget or valower 5. Then come to the second parte of the nūber .ii.C. whose valewer is 2 and denomination .C. Seke .ii. at the ryghte hande of the table, and go along vnder the denominations, towarde the lefte hande, tyll you come vnder .C. Then loke to ẏ fote of the table, and there shall you se the number of the place, that is to saye .iii. wherin you must set your digette 2. Then do so by your other two numbers that remayne, and you shall fynde 5 in the seconde place for your fyfty, and 7 in the fyrst place for your .vii. And thus maye you do with other numbers. S. Mayster I thanke you hartely, I perceaue you seke to instructe me moste playnely and breflye, and not to hyde your knowlege with subtyll wordes, as many do: For this rule is so playn that I can desyre it no playner. And though it seme somwhat longe, yet I

per=

Numeration.

perceaue it a sure waye. M. So is it, & though it be longe, yet it is nother to longe, nother to playne for yonge learners that lacke practise, for this table is in stede of a teacher, to them that lacke one. But nowe I truste, I haue sayde ynough of Numeration, which after you haue well practysed, then maye you learne forth. S. Syr, in Numeratiō I haue well practised, & am reddy to learne forth. M. That is well. But what shuld you nexte learne, can you tell? S. I remēber you said, y Additiō was next. M. Euen so, & what that is, must you fyrst knowe.

Addition

Addition is the re=
duction and bryn=
gynge of two sum=
mes or more into
one. As yf I haue
160 bokes in the la=
ten tongue, and 136
in the greke tongue, and wold know howe many they be in all. I muste
C   write

Addition.

wryte thofe .ii. nombers, one ouer an other, wrytyng the greateſt nomber hygheſt, ſo that the fyrſt figure of the one, be vnder the fyrſte figure of the other. And the ſeconde vnder the ſecond, and ſo forthe in order. When you haue ſo done, drawe vnder them a ſtrayghte lyne, then wyll they ſtand thus.

$$\begin{array}{r} 160 \\ 136 \\ \hline \end{array}$$

Now begynne at the fyrſte places towarde the ryghte hande alwayes, and put together the ii. fyrſt figures of thoſe two ſummes, and loke what cometh of them, wryte vnder them ryght vnder the lyne: As in ſayenge, 6 and 0 is 6. Wryte 6 vnder 6, as thus.

$$\begin{array}{r} 160 \\ 136 \\ \hline 6 \end{array}$$

And then go to the ſecond figures, and ſo like wayes: as in ſayenge, 3 & 6 is 9, wryte 9 vnder 6 & 3, as here you ſe. And lyke wayes do you with the figures, that ben in the thyrde place, ſayeng, 1 and 1

$$\begin{array}{r} 160 \\ 136 \\ \hline 96 \end{array}$$

be 2, wryte 2 vnder them, and then
wyll

Addition.

wyl your hole sume appere thus.

160	So that nowe you se, that
136	160 and 136, do make in all,
296	296. S. What, this is very

easy to do, me thynketh I can do it,
eue syth. There came through chepes
syde .ii. droues of cattell, in the fyrste
was 848 shepe. And in the seconde
was 186 other beastes. Those two
summes I must wryte as you taught
me thus.

Then yf I put the .ii. fyrste   848
figures together, sayenge 6   186
and 8, they make 14. That muste I
wryte vnder 6 and 8, thus.   848
M. Not so, and here are you   186
twyse deceaued Fyrst in go=   14
yng about to adde together
ii. summes of sondry thynges, which
you ought not to do, excepte you seke
onely the nomber of them, & care not
for the thynges. For the summe that
shulde resulte of that addition, shuld
be a summe nother of shepe, nor other
beastes, but a confused sume of both.

                     C.ii.   How

Addition.

How be it, somtymes you shall haue summes of dyuers denominations to be added, of whiche I wyll tell you anone: But fyrste I wyll shewe you where you were deceued in an other poynte, and that was in wrytinge 14, which came of 6 and 8, vnder 6 & 8, whiche is vnpossyble: for howe can two figures of two places, be writen vnder one fygure, and one place? S. Truth it is, but yet I dyd so vnder=stande you. M. I sayde in dede that you shuld write that vnder them, that dyd result of them bothe together, whiche sayenge is always trewe, yf that summe do not excede a dygette. But yf it be a myxte nōbre, then must you write the diget of it vnder your fygures, as I haue sayde before: But & yf it be an article, thē write o vnder thē, & kepe the article in your mynde. And therfore when you haue added your seconde figures, whiche occupy the place of tennes, you shall put that 1 therto, whiche you kepte in your mynde:

Addition.

mynde: for though it were 10 in dede, yet in ẏ place it is but as 1, bycause that euery 1 of that place is tenne, for it is the place of tennes. And in lyke maner, yf you haue in ẏ seconde place so great a nomber, that it amounteth aboue 9, then write the dyget, and reserue the article in your mynde, euer addynge it to the nexte place folowynge: and so of all other places, howe many so euer you haue. And yf you haue a myxte nomber, when you haue addyd your last figures, then write ẏ digette vnder the laste fygures, and the article in the nexte place beyonde them: So shall your nōber resultyng of addition, haue one place moze then the nombres whiche you shulde adde together. S. Now do I perceaue you, and the reason of this is (as I vnder stande) bycause that no one place can cōtayne aboue 9, which is the greatest figure that is. And then all the tennes or articles must be put to the nexte place folowyng: for euery place

C.iii.   (as I

Addition.

(as I may ſe) excedeth the other place nexte befoꝛe hym by .x. Nowe (yf it pleaſe you) I wyll returne to my example of cattell. But I remēber you ſayde I myghte not adde ſummes of ſondry thynges together, and that myght I ſe by reaſon. M. Trouth it is, yf you ſeke the ſūme of any thyng, but yf you only ſeke a ſūme, and haue no reſpecte to the thynge, then were it better to name the ſumme only without any thynge : as in ſayenge 848, without nampyng ſhepe oꝛ any thing els. And lyke wayes 186 namyng nothynge. Nowe let me ſe, how can you adde thoſe two ſūmes. S. I muſt fyꝛſt ſet them, ſo that the two fyꝛſt figures ſtande one ouer an other, and ſo then drawe a lyne vnder them bothe. And ſo lyke wayes of other figures, ſettynge all wayes the greatteſt nomber hygheſte thus.

$$\begin{array}{r} 848 \\ 186 \\ \hline \end{array}$$

Then muſt I adde 6 to 8 whiche maketh 14, that is a myxte nomber, therfoꝛe muſt I take the

Addition.

the dyget, whiche is 4, and wryte it vnder 6 and 8, kepynge the article 1 in my mynde, thus. Then do I come to the second figures, addynge them toge=ther, sayenge: 8 & 4 make 12 to whi=che I put the 1 reserued in my mynde and that maketh 13, of whiche nom=ber I write the dyget 3 vnder 8 and 4, & kepe y̆ article in my mynde, thus Then come I to the thyrde fi=gures, sayeng: 1 and 8 make 9, & 1 in my mynde maketh 10, Syr, shall I wryte the cyphar vnder 1 and 8? M. Yea. S. Then of 10 I wryte the cyphar vnder 1 and 8, and kepe y̆ article in my minde. M. What nede that, seynge there foloweth no more figures? S. Syr I hadde for=gotten, but I wyll remember better hereafter. Then seynge I am come to the laste figures, I muste write y̆ cy=phar vnder them and the ar=ticle in a farther place after the cyphar thus.

```
  848
  186
    4

  848
  186
  ───
   34

  848
  186
  ────
  1034
```

C.iiii.　　M.

Addition.

M. So now you se, that of 848 and 186 addyd together, there amounteth 1034. S. Now I thynke I am perfect in Addition. M. That wyll I proue by this example. There are .ii. armies of souldyers, in the one are 106800, & in the other 9400, Howe many are there in both armies say you? S. Fyrst I set them one ouer an other, begyn= nynge with the fyrste nombres at the ryght hande thus.

    106800
      9400

But the nether nōber wyll not matche the ouer nōber.

M. That forceth not. S. Then do I adde o to o, and there amounteth o, that muste I wryte vnder the fyrste place thus.

    106800
      9400

         0

M. Well sayde. S. Then lyke wayes in the seconde place I adde o to o, and there aryseth o, whiche I write vnder the second place thus.

    106800
      9400

        00

Then I come to the thyrd place, sayenge : 4 and 8 make 12, of which I write

the

Addition.

the dygette 2, and kepe the article 1
in my mynde, thus.   106800
Then adde I 9 to 6, whi=   9400
che maketh 15, to that I   200
adde the article 1, that was
in my mynde, and it is 16, I wryte 6
vnder 6 and 9, & kepe 1 in my mynd
thus.   106800
M. Why do you not wryte   9400
bothe figures, seynge you   6200
are come to the laste couple
of nombres? S. Nay, reason showeth
me that I must adde that article that
is in my mynde, vnto the nexte figure
of the ouer summe, though there be no
more in the nether summe. M. That
is well cõsydered, then do so. S. Then
saye I, 0 in the ouer summe and 1 in
my mynde make 1, that I wryte vn=
der 0: Then foloweth there 1 in the
ouer summe, whiche hath none to be
addyd to it, for there is none in the
nether summe, nor yet in my mynde:
therfore I thynke I muste write that
euen as it is. M. Yea. S. Then doth
C.v.   my

Addition.

my hole sume appere thus.   106800
M. Yf you marke this, you   9400
haue learned perfectly the   116200
cōmyn addition of all sum=
mes, which are of one denomination,
so that you obserue this also, that in
Addition you muste haue two nōbres
at the leaste : or els how can you saye
that you do adde? And euer let ỹ grea=
test nombre be written hyghest, for ỹ
is the beste waye, though it be not ne=
cessary. And forgette not this, that yf
you haue many nombers to adde to
gether, you shall haue oftentymes an
article of a greater valewe than 10,
somtymes 20, somtymes 30, somty=
mes more, yea paduēture 100. Ther=
fore as you dyd w̄ the article 10, so do
w̄ thē reseruing them in your mynde
& addyng to the nōbre next folowyng
so many as theyr valewer or valewe
certayne is, that is to saye : 2 for 20,
3 for 30, & so forth of other. But yf the
article be 100, then must you not adde
the article to the next figures folow=
ynge

Addition.

ynge but to the thyrde figures from them: as I wyll shewe you anone by example. And yf it chaunce the nōber to be suche, that it do comprehende two sondry articles (that is one of tennes, and an other of hundredes) then muste you reserue them bothe in your mynde, and adde the article of tēnes to the figures that folow next, and the article of hūdredes to the figure of the thyrde place from thence. Now take this example for all. I wolde adde these .xiii. summes in one, which I set after this maner. Then do I begyn & gether the sume of the fyrste figures whiche cometh to 107. For fyrste I take 9 there .x. tymes, and that is 90: then 9 and 8 is 17, that is in all 107: Of whiche summe I wryte the 7 vnder the fyrste figures, and then haue I an article of 100 in my mynde, which other

```
4889
4599
2299
3699
2399
4090
1099
3198
 299
 699
 899
 499
 389
```

Addition.

other I muste kepe in my mynde, tyll I come to the thyrde figures, whiche are in the roumes of hūdredes: or els I maye for ferae of forgettyng, write one vnder the thyrde rewe of figures makynge two lynes, as you se here done. And then muste I write the di= gettes vnder the loweste lyne, & this is the surest waye when the summe is so greate that the additiō of one rowe passeth 100. When I haue so done, I must then come to the second rowe of figures, and adde them together, which doth make 115, of which sūme I write the digette 5 vnder the same seconde rowe, and then haue I an ar= ticle remaynynge of two figures, of whiche the one (that standeth for 10) must be added to the seconde or nexte place after thē, that I dyd laste adde. And the other (that standeth for 100) must be added to the thyrde place frō thence. S. That is to saye, the fourth place from the fyrste lyne or rowe of figures. M. Euen so. And then wyll the

Addition.

the summe appere thus. Then adde the thyrde rowe of figures with the two vnitees betwene the lynes: and the summe amounteth to 50, of whiche I wryte the cyphre vnder the same thyrde rowe, and the 5 vnder the nexte figures towarde the lefte hande. Then I adde the figures of the fourth rowe with the 1 & 5, that are vnder them betwene the two lynes, and they make 29: then wryte the 9 (that is the dygette) vnder the fourth place, and the 2, that is the article beyonde it to warde the lefte hande. So those summes do make 29057. S. This semeth somewhat harde, by the reason of so many nombres together. How be it I thynke yf I do often proue euen with this same example, I shall be able to

do

```
 4889
 4599
 2299
 3699
 2399
 4099
 1099
 3198
  299
  699
  899
  499
  389
   11
   51
29057
```

Addition.

do so shortely with any other summe. M. So shall you: for it is often practise that maketh a man quycke and rype in all thynges. But bycause of suche greate summes, in whiche there may chaūce to be some errour, I wyll teache you, how you shall proue whether you haue done well or no. Sco. That were a greate helpe and ease. M. Begyn fyrst with the hyghest nōbre and then all the other orderly, and adde them together, not hauynge regarde to theyr places, but as though they were all vnities, & styll as your nōber encreaseth aboue 9, cast away 9. And then go forth, euer castynge awaye 9, as often as it amounteth therto. And so do tyll you haue gone ouer all the nombers that you entendyd fyrst to adde, and what so euer remayneth after such addition, & castynge awaye of 9, wryte it in some voyde place, by the ende of a lyne, for the better remembraunce. And then putte together the figures that resulte

Addition.

ſulte of the addition, ſtyll caſtynge a=
way 9 alſo : And then $\tilde{y}$, that remay=
neth, wꝛite at the other ende of $\tilde{y}$ lyne,
and yf thoſe two be lyke, then haue
you well done, but yf they be vnlyke,
then haue you myſſed. As foꝛ exam=
ple in this pꝛeſent ſummes : The fyꝛſt
figure of the ouer lyne is 9, let hym
go : then 8 and 8 is 16, take awaye
9, and there remayneth 7, adde to it
4 that foloweth, and that maketh 11
from whiche yf you take 9 there re=
ſteth 2. Then come to the next rowe,
whoſe fyꝛſt and ſecõde nomber are 9,
therfoꝛe ouerpaſſe them bothe ⁊ take
the 5 to the 2 which dyd remayne in
the fyꝛſte rowe, that maketh 7, putte
therto the 4 folowynge, that maketh
11, thence take 9, ⁊ there remayneth
2. Then go to the thyꝛde lyne, whoſe
two fyꝛſt nombꝛes you maye let paſſe
bycauſe they are nynes : then take the
two 2, whiche with the other 2 that
remayned in the ſeconde rowe, make
6. Then go to the fourth rowe whoſe
ii. fyꝛſt

Addition.

ii. fyrste numbers let go, and take the 6 to the 6 that remayned, and that maketh 12, take awaye 9, and there resteth 3, whiche with the 3 that is nexte, maketh 6. And so go thorough all the other numbers, you shal fynde ẏ there remayneth 5, after you haue cast awaye 9 as often as you fynde it, therfore wryte 5 at one ende of a lyne in a voyde place thus. 5————
Then gether all the fygures of the total summe which is vnder the lowest lyne, and caste awaye 9 as often as you fynde it: as thus: 7 and 5 make 12, take awaye 9 and there resteth 3, to that yf you adde the 2 that is last (for you maye let go the 9) then doth it make 5, whiche you muste write at the other ende of the lyne, that you made in the voyde place, and it wyll be thus, 5————5. And then you se ẏ those two figures be like, wherby you maye knowe that you haue done well, and so maye you proue in any other. S. Yf it please you I wyl proue in an

Addition.

in an other sume. M. With a good wyll. S. Then wyl I take one of your fourmar examples, whiche was this. Fyrst in the hyghest lyne 8 and 6 make 14, then 9 taken awaye remayne 5, to which I adde the 1 that foloweth, and that maketh 6.

```
106800
  9400
------
116200
```

Then come I to y̆ second lyne, where I fynde fyrste 4, whiche with 6 maketh 10, from that I take 9, & there resteth 1, the nexte figure is 9 and therfore I let hym alone : so I fynde 1 remaynyng, which I set at y̆ ende of a lyne thus 1——— Then I come to the totall sume, and there I muste fynde, that all the figures put togeↄther make 10, from whiche I take 9, and there resteth 1 also, which I put at the other ende of y̆ lyne thus 1——— 1 And bycause they be lyke, I know I haue well added. M. So you knowe now bothe how to adde two summes or more together. And also howe to proue whether you haue done well or no:

Addition.

no : whiche thynge also you maye do by Subtractiō. But bycause you can not yet skyll of it, I wyl let that passe tyll anone, and wyll teache you nowe how to adde summes of dyuers deno= minations : which thynge can neuer be, but when the one denominatiō is suche, that it cōtayneth the other cer= tayne tymes. And yet you shall adde them to the other, not after this sorte, as you dydde them that were of one denomination : but after such a sorte, as I wyll now showe you : That is to saye, yf you haue a summe of dyuers denominations, then loke that ye set euery denominatiō by hym selfe with some note or figure of his denomina= tion, as they be wonte to be wrytten. Then wryte your other summes so vn der that fyrst, that euery one be set vn der the other of the same denomina= tions : As for an example. Yf your de nominations be poundes, shyllyngs, & pennes, wryte poundes vnder poū= des, shyllynges vnder shyllynges, and

Addition.

¶ pennys vnder penys: And not ſhyll=
ynges vnder pennys, noʒ penys vn=
der poundes. S. Now that you haue
ſpoken it, me thinketh it neded not to
warne me of it, for it were agaynſte
reaſon ſo to confoūde ſūmes: but yet
yf you had not ſpoken of it, peraduen
ture I ſhuld haue ben deceaued in it.
M. Yf you do ſaye it is ſo playne, I
wyll ſpeake no moʒe of it, but with
an exāple make the matter to appere
euydently. Fyrſte one man oweth me
22 li. 6 s. 8 d. Another oweth me
5 li. 16 s. 6 d. And another oweth me
4 li. 3 s. I wolde knowe what this
is all together. Therfoʒe muſt I fyrſt
ſet downe my greateſt ſūme, and then
the other, euery one vnder his deno=
mination, agreynge to the greatteſt
ſumme, as thus.

Then muſt I begynne
at the ſmaleſt numbʒes
whiche muſte alwayes
be ſet nexte the ryghte
hande, and adde them

li.	s.	d.
22	6	8
5	16	6
4	3	

D.ii.   toge=

Addition.

together. And yf the summe of them wyll make one of the next denomina=tion, then must I kepe it in my mynd, tyll I come to that place: or elles for more easines write it vnder ye place be twene the double line, and vnder that fyrste place must I note the resydewe (yf there remayne any) of the same denomination, but yf there remayne none, thē nede I to write vnder it no=thynge. And this is all that you must marke in this addition: for all other thynges are lyke to the other maner of addition, before mentioned. Ther=fore the cheyfest poynte of this addi=tion is, to knowe the valewes of cō=men coynes, and rated summes: As how many shyllynges be in a pound, how many pennys in a shyllynge. Of whiche and of other lyke thynges I wyll instructe you here after, in tea=chynge of Reductiō. But now I may not disturbe your witte from ye thyng that we are about. Therfore lette vs retourne to that fourmer example,

which

Addition.

which I proposed of .iii. detters, whi=
che summes when I had set orderly,
they stode thus with a
double line vnder thē.
Then to adde them vn
to one sūme, must I be=
gyn at the ryght hand,
where the smalest deno

li.	s.	d.
22	6	8
5	16	6
4	3	

mination is, and adde them together
fyrste sayenge, 6 & 8 make 14, nowe
seynge these 14 are pennes, and that
12 d. make one shyllynge (which is y͏̆
nexte valewer) I take away 12 from
14, & there resteth 2, whiche I write
vnder the pennes, and for the other
12 (whiche make 1 shyllyng) I write
1 vnder the tytle of shyllynges, thus.
Then do I adde all the
shyllynges together, &
fynde them 25, to which
I adde the 1 betwene y͏̆
two lines, and that ma=
keth 26 : but bycause y͏̆
20 s. do make 1 li. I

li.	s.	d.
22	6	8
5	16	6
4	3	
	1	
		2

take away 20 from 26, and for that
                 D.iii.   20

Addition.

20 I write 1 vnder the poundes, be=
twene the two lynes: and the other 6
(that remayneth) I write vnder the
shyllynges thus.
Then come I to the poū=
des addynge them all to=
gether, and fynde them
to be 31, thereto I adde
the 1 betwene the .ii. ly=
nes, and that maketh 32
whiche summe I write downe hole,
because there is no greatter denomi=
nation. And then my hole summe ap=
pereth thus.
So is my totall summe
32 li. 6 s. 2 d. And this
may you proue in an o=
ther lyke summe. S. Then
wyll I caste the hole
charge of one monthes

li.	s.	d.
22	6	8
5	16	6
4	3	
1	1	
	6	2

li.	s.	d.
22	6	8
5	16	6
4	3	
1	1	
32	6	2

cōmens at Orforde, with battelynge
also. M. Go to, let me se how you can
do. S. One wekes cōmens was 11 d.
ob. q̈. q. and my battellynge that weke
was 2 d. q̈. q. The second wekes cō=
mens

Addition.

mens was 12 d. & my battelyng 3 d. The thyrde wekes cōmens 10 d. ob. & my battelyng 2 d. q. c. The fourth wekes cōmens 11 d. q̈. and my batte=lynge 1 d. ob. c. These .iiii. summes wolde I adde into one hole summe, & therfore I wyll set them one ouer another thus.

But I had forgottē, I shuld haue set the greatest sume hy=ghest. M. So is it cō menly best, how be it here it forceth not, & in such sūmes as this is, that go by order of wekes, dayes, or yea=res, it is better to kepe

	d.		
11	ob.	q̈.	
2		q̈.	q.
12			
3			
10	ob.		
2		q.	c.
11		q̈.	
1	ob.		c.

that order then to alter them, and to sette the greatest nombre hyghest, for that serueth for suche summes as go not by order. S. Then yf I haue sette them well ynough, I wyll begyn to adde them thus. Fyrst of the smaleste valewers at the ryght hand, whiche
D.iiii.   are

Addition.

are called ceys, I fynde 2, & seynge ye 2 ceys do make 1 q. I wyll write nothynge vnder the ceys, but wyll write 1 q. for the 2 ceys vnder the kewes betwene the lynes, after this maner. Then come I to the next valewers where I finde 2 q. & to them I adde the q. that is betwene the lynes and so are they 3 q. but bycause 2 q. maketh one q̈. I wryte one q̈ vnder the farthynges betwene ye lynes, and the q that remayneth muste I wryte betwene the nethermost lyne, vnder the kewes thus.

Then come I to the farthynges, where I fynde 3, and the other q̈ that is betwene the lynes maketh 4 q̈.

	d.		
11 ob.	q̈.		
2		q̈.	q.
12			
3			
10 ob.			
2		q.	c.
11	q̈.		
1 ob.			c.
1		q.	
1			

	d.		
11 ob.	q̈		
2		q̈	q.
12			
3			
10 ob.			
2		q.	c.
11		q̈	
1 ob.			c.
		q̈.	q.
		q.	

Addition.

and bycause 4 q̃ make iuste 1 d. I shall write nothynge vnder the far=thynges, but must write 1 vnder the penēs betwene the lynes. Then must I adde the halfe pennes together, of whiche there are 3, but seynge that 2 ob. make 1 d. I muste write 1 vnder the pēnes betwene the lines, but how shall I do it, for there is 1 alredy? M. Haue you forgottē how I dyd in ad=dition of the greate sūme before? you must set it vnder ẙ other, so shall they bothe stande for 2 : for yf you shulde set it before or behynde the other, they shulde make 11. S. I remēber it now, and I perceaue the reason. Then I wyll write 1 ob. vnder ẙ halfe pēnes, and for the other 2 halfe pēnes whi=che make 1 d. I write 1 vnder the pēnes. Then come I to the pennes, & fynde that there are of them 52, then put I to them the 2 betwene the ly=nes, and that maketh 54, whiche a=moūteth to 4 s̃. 6 d. the 6 d. I must write vnder the pēnes, and the 4 s̃.

D.v. I

Addition.

I muste sette (I suppose) farther to‑
warde the lefte hande by them selfe.
M. Euen so. S. Then appereth all my
addition thus.

And the summe is
4 s. 6 d. ob. q. M.
Nowe haue you
done this well.
But tell me, why
dyd you writ kewe,
cee, thus q. c. and
not rather thus q^c,
as the fashyon is?
Scol. Bycause I
thought it was ye
best waye for dewe
gatherynge of euery denominatiō by
hym self. M. So was it in dede. Well
now, can you tell howe to proue this
addition (and suche other lyke of dy‑
uers denominations) & to trye whe‑
ther you haue done well or no? S. I
wolde I coulde. M. That shall you
do by this meanes : fyrste as you dyd
begyn to adde, so reken agayn eue‑
ry de‑

	d.		
	11 ob.	q̈.	
	2	q̈.	q.
	12		
	3		
	10 ob.		
	2		q. c.
	11	q̈.	
	1 ob.		c.
	1	q̈.	q.
s.	1		
4	6 ob.	q.	

Addition.

ry denomination by it selfe, and whē you kynde so many small, y̆ do make any other denomination, let them go and kepe in mynde only the resydewe that wyll make no greatter denomi= natiō, and loke whether there be any such lyke valewe vnder y̆ nether line, and yf there be, you haue well done, and so go forth from one denomina= tion to an other vnto the ende. But here must you note that in gathering of the sūmes you must recken those fi gures that are wrytten betwene the lynes with them that are written a= boue them, as for an example. J wyll examyne that sūme, that J dyd last adde, whiche stode thus.

li.	s.	d.
22	6	8
5	16	6
4	3	
1	1	
32	6	2

Fyrste J kynde 6 and 8 which maketh 14, from which J take 12, bycau se it maketh one of the nexte denomination, and there remay neth 2, and vnder that place J se a lyke figure, therefore J knowe that to be

Addition.

be well done. Then come I to the s. where I fynde 1, 3, 16, and 6, that maketh 26, I cast away 20, for they make an other denomination, that is to say, poundes, & the 6 which remayneth is lyke to the 6 that is written vnder them, benethe the lowest lyne, therefore that is well done also. And thence I go to poũdes, where I fynd 1, 4, 5, 22, yͭ is 32, to whiche sũme agreeth another like vnder it, therefore I iudge all well done. S. I perceaue reasõ in this probatiõ, now wyl I attempt the same in yͤ sũme yͭ I dyd adde, which when I had ended adding, stode thus. Fyrst amongest the cees I fynd but 2, whiche make 1 q. euen therfore there must nothyng be vnder the lyne for them: and amongest the kewes, I fynde 3 of

	d.		
11	ob.	q̈.	
2		q̈.	q.
12			
3			
10	ob.		
2			q. c.
11		q̈.	
1	ob.		c.
1		q̈.	q.
s. 1			
4 6	ob.		q.

Addition.

3, of whiche 2 make 1 q̄. therfore I let them go, and the 1 q. that is lefte hath an other lyke vnder his place, therfore that is well done. Then the farthynges are iust 4, which make 1 d̄. and therefore I let them go. Amongest the halfe pēnes there is one odde for 2 must I cast away, bycause they make 1 d̄. and vnto it answereth a lyke summe vnder it. The pennes are 54, from whiche I take awaye 48, y̌ make 4 ʃ. and the 6 remaynynge agre to a lyke figure sette vnder them. And laste of all remayneth the 4 ʃ. which the abiected pēnes dyd make, so I perceaue that I haue well done. Now this wyll I not forgette, but wyll this examinatiō serue in all addition? M. It serueth for all additiō of sundry denominatiōs, yf the addition be made with two lynes (as these were) els it wyll not serue, bycause y̌ those summes which are here addyd betwene the lynes in addition, by one lyne are vnderstande and not written, but I

Addition.

but I lette that wayes passe, bycause as it is commyn, so is it more deceaueable then this wayes, namely yf a mans memory be other dulle, other troubled. S. yet it were good to know that wayes also. M. Yf you desyre to knowe it, this it is in fewe wordes: do euery thynge as you dyd in this sorte, saue that where you made here two lynes, you shall make there but one, and those summes that you dyd here wryte betwene ye lynes, you must kepe in your memory and adde them (as you dyd here) eche one when you come to his place. S. Then they dyffer not but in this, that this addition with two lynes leueth nothynge to memory, but wryteth downe all: and the other waye comptteth certayne nombers to memory, as you taught me in the fyrste examples of addition of small summes of one denomination. But what yf a man vse it (as you say men do comenly) how shall it be examined? M. Seynge you are so desyrous

Addition.

rous of it I wyll shewe you bothe an example of the addition, and also the maner to examyne it. I propose these iii. summes to be added and I gather fyrste the penes (as I dyd in the other sorte) and I fynd of them 8, 3, 9, that is

li.	s.	d.
12	8	9
3	4	3
3	6	8

20, of whiche summe I bate awaye 12, which make 1 s. and kepe that 1 in my mynde, and the rest that is 8, I write vnder the pennes. Then do I adde the shyllynges together, & fynd of them 6, 7, 8, that is 21. wherof I bate 20, that make 1 li. which I kepe in my mynde, and to the other 1 that remayneth, I adde that 1 that came of the pennes, and was in my mynde, whiche make 2 : & them I write vnder the shyllynges. Then do I recken the poundes together 3, 6,

li.	s.	d.
12	8	9
6	4	3
3	6	8
22	2	8

12, that is 21, and to them I adde ye 1 in my mynde that remayned of the
shyllyn=

Addition.

shyllynges, which make 22, them do I wryte vnder the poundes : and then my summe totall appereth to be 22 li. 2 s. 8 d. Now to exampne this sũme (and all lyke) you shall do thus. Fyrst begyn at the lefte hand with the poũ= des, & take from them, that are aboue the lyne 9 as often as you can, then that $\overset{t}{y}$ remayneth shall you double, then ioyne it with the shyllinges, and take awaye 9 from that, as often as you can, and what so euer remayneth you shall take for it 3 tymes so moch and putte to the pennes : then take frõ all that sũme 9 as often as you can, & what so remayneth after you haue withdrawen 9 as often as you can, wryte that at the ende of a lyne, as I taughte you in other addition. And then come to the sũme vnder the lyne, begynnynge with the poundes, and do euen as you dyd with the summes aboue the lyne, tyll you come to your pennes, and yf the figure of the sũme that remayneth after castynge away

9 (as

Addition.

9 (as often as you can) do agre with the other that remayned before of the other summe, whiche you dyd wryte at the ende of the lyne, then haue you done well, els not: ¶ for an example I wyll examyne that last sume which was thus.

Fyrst I shall begynne at the lefte hande with the poundes, puttynge them together, whiche make 21, in which sum

li.	s.	d.
12	8	9
6	7	3
3	6	8
22	2	8

I fynde 9 twyse, for twyse 9 is 18, ẏ I deducte ¶ there remayneth 3, that 3 muste I double (as I sayd) bycause it is the remayner of the poūdes, and it wyll be 6. Then gether I the sume of the pennes, whiche is 21, to the whiche I adde the fore sayde 6, and then is it 27, wherin I fynde 9 thre tymes, and there remayneth nothing This remayner shulde I take .iii. ty= mes, but .iii. tymes nothynge is no= thyng: therfore in this place is there nothyng lefte to be added to ẏ pēnes,
                                                                                 ¶.   ther=

Addition.

therfore muste I take the summe of pēnes aboue, which is 20, frō thence yf I take 9 twyse, there remayneth but 2, whiche I put at the ende of a lyne, thus. 2——— Then I come to the poundes of the vnder nōber or totall summe, and there I fynde 22, frō whiche I take awaye 9 twyse, and there remayneth 4, that 4 I double and it is 8, then do I adde that 8 to the shyllynges, and it maketh 10, frō whiche I withdrawe 9, and there resteth 1, then do I take that 1 thre tymes, and it maketh 3, whiche I adde to the 8 d. & it maketh 11, frō which yf I bate 9, there resteth 2, which is equall to the nōber noted at the ende of the lyne, and therby I perceaue ẏ I haue done well. S. But I do not se the reason of this. M. No, no more do you of many thynges els, but hereafter wyll I shewe you the reasons of all Arithmeticall operatiōs, for this I iuge to be ẏ best trade of teaching, fyrst by summe brefe preceptes to en=
structe

Addition.

structe a learner sumwhat in the vse of the arte, before he learne y̑ reasons of the arte, and then maye you after=warde more soner make hym to per=ceaue the reasons : for harde it is for to occupye a yonge learned wytte, w̄ both the arte and the reasons of it all at ones : how be it he shall neuer be cōnyng in dede in an arte, y̑ knoweth not the reason of euery thynge tou=chynge it. S. Yet at the least, I praye you show me why dyd you write your nōbre that remayneth (after you had withdrawen al the nynes) at the ende of a lyne, for I sawe no reason why y̑ lyne dyd serue? M. Dyd you euer marke a carpenter when he wrought? S. Yea, many tymes. M. And haue you not sene hym when he hath taken measure of a borde, that he hath pric=ked it, and hath with a twyche of his hande drawen a lyne from the pricke that he made? S. Yes I haue marked that, and haue sene some make .iii. or iiii. lynes by the pricke, and some also

E.ii.   haue

Addition.

haue I sene make a crosse by it, but that I perceaued was for the easy findynge of theyr pricke. M. And euen so is this lyne, for the easy fyndynge of your remayner, and therfore some do make a crosse thus.
And set the one remayner a=
boue the crosse, and the other
vnder the nether parte of the crosse, as yf I shulde set my two remayners thus.　　　　　2
But there is an other sorte of
profe of addition, to whiche 　2
the crosse serueth more meter. And $\tilde{y}$ is, when the addition is of dyuers denominations, and I wolde examyne euery denomination by it selfe, which wayes though it be not moch vnlyke to the fyrste profe, that I brought of such dyuers summes, yet I wyll de= clare it, least you shulde thynke that I wolde hyde it from you: you muste make so many lines in your crosse as you haue sondry denominations, as yf you haue but 2 denominations,
then

Addition.

then you maye make it thus:
that the ouer parte and the
nether parte maye serue for
one denomination, and the two sydes
for the other. And yf you haue 3 de=
nominations (as poūdes, shyllinges,
and pennes, then must you make thre
lynes, thus.

The vpryght lyne may serue
for the poundes, and the hy=
ghest thwarte lyne for shyl=
lynges, and the lowest for pennes, as
for example, I wyll take a sūme thus
added, for the proue of the whiche,
bycause it contayneth thre denomina
tions, I must make a crosse of 3 ly=
nes thus.

```
    li.   s.   d.
    16   12    5
    12    8    1
     9    2    7
    ──────────
    38    3    1
```

Then I reken fyrste at the ryghte
hande the pennes 7, 1, 5, make 13,
from whiche I take 12 for the nexte

C.iii.   de=

Addition.

denomination, that is to saye a shyl=
lynge, and there resteth 1, whiche I
must write at one ende of the nether
thwarte lyne. Then I gether ẙ sum̄e
of the shyllynges 2, 8, 12, which make
22, to them I put 1 that I toke of the
pennes, & that maketh 23, from those
I take 20, the quantite of the nexte
greater denomination (that is to say
a pound) and there resteth 3, whiche
I write at the ende of ẙ ouer thwarte
lyne. Then I adde together the pou̅=
des, 9, 12, 16, whiche make 37, to thē
I adde the 1 that came of the shyllyn
ges, and then there is 38, wherin I
fynde 4 tymes 9 and 2 ouer : that 2
I write on the vpryght lyne. Then
I come to the totall summe, and exa=
myne it begynnynge at the pennes,
where I fynde but 1, & can not take
9 from hym, therfore I set hym at the
other ende of the nether thwart lyne.
Then come I to ẙ shyllynges where
I fynde only 3, whiche bycause it is
lesse then 9, I set it at the other ende
of

Addition. 36

of the lyne of ſhyllynges, that is the ouerthwarte lyne. Then of the 38 li. I take out 4 tymes 9 (whiche is 36) & there remayneth 2, whiche I wryte vnder the vpryghte lyne, then I conſyder euery nomber, comparynge it to the nomber that is agaynſt it : and bycauſe I kynde them to be euer one lyke his matche, I knowe that I haue well done. S. This croſſe I pceaue doth ſerue for thoſe thre denominations, poundes, ſhyllynges, pennes : but what yf I hadde, ob, ḡ. q. and c? M. You thynke you be at Oxforde ſtyll, you bꝛynge forthe ſo faſte your q, and c. Theſe thre lynes (as I haue ſayde) doth ſerue for thre denominations, ſuch as they be : as here they do ſerue for poūdes, ſhyllinges, & pēnes : but yf you haue no poundes in your ſumme, then maye they ſerue for ſhyllynges, pennes, and halfe pēnes, yea for ḡ, q, and c, yf you haue no greater denomination, ſo that you remember that the vpꝛyght lyne ſerueth for the

E.iiii.   grea=

Addition.

greateſt denomination, & the hygheſt thwart lyne for the nexte, & the loweſt for ẏ leaſte. And ſo yf you haue foure denominations, you muſt make your croſſe with ſo many lynes,

And yf that your ſumme be of moꝛe denominations, make ſo many lynes in your croſſe.

And thus wyll I make an en=de of Addition.

⁋ Examples of addition.

li.	s.		li.	s.	d.
262587	6		340	17	10
41635	12		28	6	8
28124	2		13	13	4
471	4		382	17	10
332818	4				

The proſes

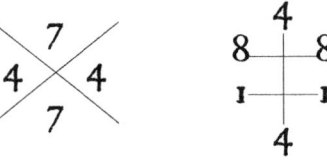

Addition.

❡ An other example.

li.	s.	d.	q.
22		6	2
2	3	4	
	10	2	3
24	14	1	1

```
    6
5 ——— 5
1 ——— 1
1 ——— 1
    6
```

❡ Subtraction. Sco.

Hen haue I learned the two fyrste kyndes of Arithmetike, nowe (as I remēber) doth folowe Subtractiō, whose name me thyn=keth doth soūde contrary to Additiō. M. So is it in dede: for as Addition increaseth one grosse sume, by bryn=gynge many in to one: so contrarye wayes Subtraction dimynyssheth a grosse summe by withdrawynge of o=ther from it: so that Subtraction or rebatyng is nothyng els, but an arte to withdrawe and abate one summe frō an other, that the remayner maye appere. S. What call you the remay=
E.v.   ner?

Subtraction.

ner? M. You maye perceaue by the name. S. So me thynketh, but yet it is good to aſke the trouth of all ſuch thynges, leaſt in truſting to my owne coniecture I be deceaued. M. So is it the fureſt waye. And as I ſe cauſe, I wyll ſtyl declare thynges vnto you ſo playnely that you ſhall not nede to doubte. How be it, yf I do ouer paſſe it ſomtymes (as the maner of men is to forgette ẏ ſmall knowlege of them to whome they ſpeake) then do you put me in remembraunce your ſelfe, & that waye is fureſte. And as for this worde that you laſte aſked me, take *Remay* you this deſcription: The remayner *ner* is a ſumme lefte after dewe workyng whiche declareth the exceſſe or diffe= rence of the two other nombers: as yf I wolde deducte 14 out of 18, there ſhulde remayne 4, whiche is called the remayner, and is ẏ difference be= twene thoſe two nombers 14 and 18. S. I perceaue then what ſubtraction is. Nowe reſteth to knowe the arte to
worke

Subtraction.

worke by it. M. That shall you do by this meanes: fyrste you must consy=
der, that yf you shulde go aboute to rebate, you muste haue two sundry summes proposed, the fyrste which is your grosse summe or sūme total, and it must be set hyghest, and then the re=
batement or sūme to be withdrawen, whiche muste be sette vnder the fyrste (whether it be in one parcel or in ma=
ny) and that so, that the fyrste figu=
res be one iuste ouer an other, and so the seconde, and thyrde, and all other folowynge, as you dyd in Addition: then shall you drawe vnder them a lyne, and so are your sūmes dewly set to begyn your workynge. Then be=
gyn you at the righte hande (as you dyd in Addytion) and withdrawe the nether nomber out of the hygher, and yf there remayne any thynge, wryte that right vnder them beneth y̆ lyne, and yf there remayne nothynge (by reason that the two fygures were e=
quall) then wryte vnder them a cy=
phre

Subtraction.

phre of nought. And so do you with all the other figures, euer more abatinge the nether out of the hygher, and wryte vnder them the remayner styl, tyll you come to the ende. And so wyll there appere vnder ẏ lyne what remayneth of your grosse sūme, after you haue deducted ẏ other sūme frō it (as in this example) I receaued of your father 48 s̸. of whiche I haue layde out for you 36 s̸, nowe wold I knowe what doth remayne, and therfore I set my nombers thus in order, Fyrste I wryte the greatest summe, & vnder hym the lesser, so that the figures at the ryght syde be euen one vnder an other, and so the other thus.

s̸.	Then do I rebate 6 out of 8,
48	and there resteth 2, whiche I
36	write vnder them ryght beneth
	the lyne, thus.

Then I go to the seconde figures, and do rebate 3 out of 4, where there remayneth 1 whiche I wryte vnder them ryghte,

| s̸. |
| 48 |
| 36 |
| 2 |

and

Subtraction.

and then the hole sūme and operatiō
appereth thus.   s.
Whereby it appereth, that yf I   48
withdrawe 36 out of 48, there   36
remayneth 12. S. Now wyll I   ──
proue in a greater sūme. And I wyll 12
subtracte 2367924 out of 3468946,
Those sūmes I sette in order thus.

3468946   Then do I begynne at
2367924   the ryght syde, and de=
────   ducte 4 out of 6, and
there resteth 2, whiche I wryte vn=
der them. Then go I to the seconde
figures, and withdrawe 2 out of 4,
and there remayne 2, whiche I sette
vnder them also : then I take 9 out
of 9, and there resteth 0, whiche I
write vnder them, for you say, that yf
the figures be equal, so that nothyng
remayne, I must write this cyphar 0
vnder them. M. It was well remem=
bred, now go forth. S. Then come I
to the .iiii. place and drawe 7 out of
8, and there remayneth 1, whiche I
wryte vnder them also : then in the .v.
place

Subtraction.

place I take 6 from 6, and there re= steth nowght, for I wryte vnder thē a cyphre 0: then in the .vi. place 3 re bated from 4, remayneth 1, whiche I wryte vnder them: and lyke wyse in the .vii. and laste place 2 taken from 3 there is lefte 1, whiche I wryte vn= der thē: so haue I done my hole wor= kynge, and my sūmes appere thus:

```
3468946
2367924
———————
1101022
```

whereby I se, that yf I rebate 2367924 oute of 3468946, there re= mayneth 1101022. M. This is well done, and that you may be sure to perceaue fully the arte of subtraction, lette me se howe can you subtract 52984732 out of 8250003456 S. Fyrste I sette downe the greateste sūme, and then I wryte vnder it the lesser nomber, begynnynge at yͤ ryght syde, and then my figures wyll stande thus.

```
8250003456
  52984732
——————————
```

Then take I 2 frō 6, and the reste is 4, whiche I wryte vnder them, then

do

Subtraction.

do I withdrawe 3 from 5, and there remayne 2, whiche I wryte vnder them : then take I 7 out of 4, but that can I not, what shall I now do? M. Marke well what I shall tell you now, how you shall do in this case, & in all other lyke. Yf any figure of the nether summe be greater then the fi=gure of the sūme that is ouer hym, so that it can not be taken out of the fi=gure ouer hym, then must you put 10 to the ouer figure, and then consyder how moch it is : and out of that hole summe withdrawe the nether figure, and write the reste vnder them. Can you remēbre this? S. yes that I trust I shal. Now then in my exāple where I shulde haue taken 7 out of 4, & coulde not, I put 10 to that 4, which maketh 14, from it I take awaye 7, & there resteth 7 also, which I write vnder them. M. So haue you done well, but now must you marke an o=ther thyng also : that whē so euer you do so put 10 to any figure of the ouer nom=

Subtraction.

nomber, you muste adde one styll to the figure or place ẏ foloweth nexte in the nether lyne, as in this example there foloweth 4, to which you must put 1 and make hym 5, and then go on, as I haue taught you. S. Then shall I saye 4 and 1 (whiche I must put to hym for the 10, that I added to 4 before) make 5, whiche I shuld take out of 3, but that can not be, therfore must I put to it also 10 and then it wyll be 13, frō whiche I take 5, and there resteth 8 to be written vnder them: and bycause of that 10 added to the 3, I muste adde 1 to 8 that foloweth in the nether lyne, and maketh 9, whiche I shulde take out of 0 and can not, therfore I put therto 10 and that maketh 10, from 10 I take 9, and there remayneth 1, which I wryte vnder them. Then do I adde 1 lykewyse to the nexte figure beneth which is 9, and that maketh 10, that 10 shulde I take out of the figure aboue, but I can not, for it is 0, therfore

Subtraction.

foꝛe I put 10 to it, and ſo take I 10 out of 10, and there reſtyth 0 to be wꝛytten vnder them : then come I to the nexte fygure, whiche is 2, and to hym do I adde 1, whiche makyth 3, that 3 I can not take out of nawght, therfoꝛe of that nawght I make 10, and thenſe do I take 3, ſo remaynyth there 7 to be wꝛytten vnder them : lyke wyſe do I take 1 to 5 that folo= weth, and then is it 6, that wolde I take out of 5 and can not, therfoꝛe I take 10 to that 5, and make it 15, from whiche I rebate 6, whiche I wꝛite vnder them : nowe haue I ſpente all the nether figures, and what ſhall I do moꝛe? M. you ſhulde haue addyd 1 to the nexte fygure folowyng (yf ther had ben any) bycauſe you addyd 10 to the laſte fygure befoꝛe of the ouer lyne. But ſeynge there is no fygure folowynge, you muſt adde that 1 to the place folowyng, and then deducte that 1 from the nombꝛe aboue. Scol. Then ſhall I ſaye bycauſe I boꝛo=

F   wyd

Subtraction.

wyd 10 to the ouer 5, I must put 1 in the nexte place beneth, that is vnder 2, then must I subtracte that 1 from 2, and there restyth 1 to be written vnder that 2 in the 9 place: nowe I haue no more to subtract, for there is neuer a fygure remaynyng beneth nother yet any vnyte to be added, bycause I borowed not 10 to the figure last before, and yet is there 8 remaynynge in the ouer lyne, whiche (I thynke by reason) shulde be set at the ende of ỹ figures in the lowest rowe, whiche is vnder the lyne, for bycause there was nothyng taken from it. M. That is well considered, and reason teacheth so in dede: how be it (as I sayde before) I wyll omytte the reasons tyll another more conueniente tyme, and wyl onely at this tyme teache you the practise of the arte, that you may exercyse your selfe sumwhat in it, in the meane tyme, by proper & wytty questiōs, as I wyll teache you some, before we departe. And when you

Subtraction.

you haue well occupyed your mynde and quyckened your witte in it, then wyll I afterwarde at your returne, enstructe you in the reasōs of all this workynge, that then you may worke perfectely, when you se the reasons in euery thynge, why you shulde so do. S. Syr well I maye praye to God to recompence this your goodnes: but surely I can not do it with any temporall benefyte, nother any other mā in my behalfe: for what treasure is there to be compared to the ryches & treasure of lernynge? But syr (I beseche you) shall I alwayes when any nomber so remayneth alone (as this 8 dyd) wryte hym vnder the lyne, strayght agaynst his owne place? M. yea what elles, whether they be one or many: and this well remēbred, you haue learned subtraction. How be it, bycause of certeyne thynges y̑ myght deceue you, yf you dyd not take good hede to your workynge. I wyll propose to you an other example of many

Subtraction.

ny nobers to be subtracted, as thus. I receyued of a frende of myne to kepe 2869 crownes, whiche at one tyme I delyuered hym agayne 500, at an other tyme 368, and at an other tyme 440, & an other tyme 80, & an other tyme 64. Now wolde I know how many doth rest behynde. Therfore fyrst I set downe my grosse sume & vnderneth hym I set all y͘ parcels, thus. And vnder them a double lyne.

Then fyrste I begynne at the fyrst place, and gather together the sume of all those lynes (saue the ouermoste) in theyr fyrste figures, and so do

```
2869
 500 ⎤
 368 ⎥
 440 ⎬
  80 ⎥
  64 ⎦
─────
```

I with all the figures of the seconde place, and so forth, as I dyd in addition, saue that I leue out the hyghest rowe of nöbres: and that sume so gathered betwene the lynes, do I subtract out of the hyghest rowe of nombres, and y͘ remayner do I set vnder the

Subtraction.

the nethermoſt lyne, as foꝛ example: I ſet the ſūmes as be= foꝛe: then do I gather the fyꝛſt fygures toge= ther, where I fynd but 4 and 8 that make 12 (foꝛ .iii. ciphers encreaſe no ſumme in addition, as you learned befoꝛe) of ẏ 12 therfoꝛe do I

```
2869
 500
 368
 440
  80
  64
————
1452
1417
```

wꝛite the dyget 2, ⁊ kepe the article in my mynde tyll I come to the ſeconde places, where I fynde 6, 8, 4, 6, that make 24, to them I put the article in my mynde, and it is 25, of whiche I wꝛyte 5 vnder the ſeconde place, ⁊ kepe the dyget 2 in my mynde foꝛ the iii. place, where I fynde 4, 3, 5, that make 12 to the whiche I adde the 2 in my mynde, and that maketh 14, therof I wꝛite the 4 vnder the thyꝛd place, and bycauſe there remayneth no moꝛe figures to be added, I wꝛyte the diget 1 in ẏ fourth place, as you ſe in the exāple. Then come I to ſub=

F.iii. tra=

Subtraction.

tractynge of this summe, and I shall subtracte the summe betwene the lynes, from the ouermost sume, sayeng: 2 from 9 remayne 7, to be wrytten vnder them, beneth the lowest lyne: then in the seconde place I take 5 from 6, and there resteth 1 to be wrytten vnder them. Then in the thyrde place, 4 from 8, resteth 4. In the .v. place 1 from 2, remayneth 1. And thus I see that after those 5 sumes are subtracted from 2, 8, 6, 9, the remayner is 1417. S. This I perceaue, but is there no shorter waye and more spedier? M. yes, when you are a whyle exercysed in it: for you maye as faste as you can gather the nombers toge ther, withdrawe them out of the hyghest sume: yf so be it, that all the par cels which you do gather do not excede 9: but and yf they excede 9 then must you subtracte only the digette, that is in it, & reserue the article tyll the nexte place, where you shall adde it with the other fygures, and so subtract

Subtraction.

tracte y̑ hole out of the figure aboue them : but & yf in this place the sũme of the parcelles do excede 9, then (as I sayde before) subtracte the digette only, and reserue y̑ article to the next place, and so styll go forthe, tyll you haue ended your workynge : as for example in the laste summes proposed, I gather fyrst in the fyrste place 4 & 8 y̑ maketh 12, of which I deduct the dygette 2 out of 9, and write vnder the remayner, whiche is 7, & the article 1 I kepe in my mynde. Then in y̑ seconde places I gather the parcels 6, 8, 4, 6, whiche amount to 24, to y̑ I adde the article 1 (whiche I haue in my mynde) and then is it 25, then do I take 5 (that is y̑ digette in this nomber) from 6, that is in the second place of the hyghest summe, and there remayneth but 1, to be written vnder them, and now do I kepe the article 2 in my mynde styll. Then in y̑ thyrd place 4, 3, 5, maketh 12, and the article 2 in my mynde, maketh 14, then

F.iiii.   take

Subtraction.

take 3 4 (whiche is the dyget) from 8, that is ouer them, and there resteth 4, whiche I wryte vnder them. Then haue I the article 1 yet in my mynde, whiche I shulde adde to the parcels nexte folowyng, but seynge there is no nomber folowynge, I take that dyget alone, and deducte hym out of the nexte summe aboue, whiche is 2, and then is the remayner 1, which I wryte in the fourth place vnder 2. Lo nowe haue you a shorter waye. S. I lyke bothe wayes well, & I perceaue bothe well: yet as in the one, the wor= kynge semeth somwhat longe, so in ẙ other it leueth very much (me semeth) to remembraunce, and therfore maye cause errour quyckly, excepte a man haue a quycke and exercised remē= braunce. M. What, wolde you then haue suche a waye, that shulde not be so longe as the one, nor so shorte as the other? S. Yea, yf there were any suche. M. Then do thus: styll as you gather your parcelles when they ex=
cede

Subtraction.

cede a dyget, and maketh 10 or more, take the article & write hym betwene ii. lynes (as in the fyrste example), vnder the nexte place towarde the lefte hande, and then deducte the dygette from the figure that is ouer hym, and wryte the remayner. And then when you gather the nexte parcelles, you shall adde to them the figure that is vnder them, betwene the .ii. lynes : & yf it excede 9, do as I sayde before, write the article vnder the next place betwene the lynes, and subtracte the digette from the figure that is ouer those parcelles : and yf that all ỹ parcelles together, and the nomber betwene the lynes do make but a diget, then deducte it holy from the figure aboue : as in this example. I wolde subtracte out of 40308964, these. iii. parcels 20003428. 10002342. 10101461, therfore I sette them fyrst in order dewe,

```
40308964
20003428
10002342
10101461
————————
```

F.v.   and

Subtraction.

and then I gather the parcels of the fyrste place, whiche are 8, 2, 1, that is 11, of which I take away the article, and set hym vnder the seconde place betwene the lynes, and the dyget 1 y̌ remayneth I deducte out of 4, and there resteth 3, to be wrypten vnder y̌ fyrst place benethe the lowefte lyne. Then come I to the seconde place, & gather the parcels of it 6, 4, 2, & the 1 betwene the lynes whiche make 13, of whiche I take the article and sette hym vnder the thyrde place, betwene the lynes, and the dyget 3 I wryte vnder the second place benethe y̌ lowest lyne. Then in the thyrd place I fynd 4, 3, 4, which with the 1 betwene the lynes do make 12, therfore I wryte the article agayne vnder the fourthe place, and the dyget 2 I take from 9 & there remayneth 7, whiche I wryte vnder them, benethe the lowest lyne. And then come I to the .iiii. place, where I gather 1, 2, 3, and the 1 be=
twene y̌ lynes, that maketh 7, which
by=

Subtraction.

bycause it is but a diget I plucke frō 8, and the remayner is 1, and must be written vnder them, in the .iiii. place. Then come I to the .v. place, where are onely .iii. cyphars, whiche make nothynge, then shulde I take that (y is to saye, nothyng) from the figure ouer them, whiche is also a cyphre, therfore I muste saye thus: yf I take nought from nought, there remayneth nought, so muste I wryte a cypher vnder them. Then in the .vi. place, I fynde but 1, whiche I take out of 3 ouer hym, and the remayner is 2, that must be written benethe y lowest lyne in the .vi. place. So go I to the .vii. place where I fynde onely cyphers, & in y grosse sume ouer them a cypher also, therfore must I wryte theyr remayner (whiche is nothynge) with a cypher also. Then in the .viii. and laste place, I gather 1, 1, 2, that make 4, which yf I take out of y 4 that is ouer them, there wyl nothyng remayne, & that must be noted with a
cyphre

Subtraction.

cyphre benethe the lowest lyne : as I haue often sayd. And thus haue I ended my worke, and the figures stande thus.

S. Syr I remembre, you taught me that cyphers shulde not come in ẏ last place, for bycause they ser= ve onely to encrease

```
40308964
20003428 ⎤
10002342 ⎥
10101461 ⎥
     111 ⎦
─────────
00201733
```

the valewe of other figures whiche folowe them, and serue not for those figures that go before them : and now in your example you haue set ii. cyphers in the two laste places.

M. I cōmende you for your remem= braunce, and truth it is, I shulde not haue set them here, but only bycause that I wolde make you playnely to perceaue the arte of subtraction, ther fore seynge you do nowe perceaue it, when so euer you shulde write downe a cypher, loke whether any other fy= gures be yet behynd, and yf not, then let go ẏ cypher also, for it nedeth not

to

Subtraction. 41

to write him in any later places, wher no other fygure doth folow, except it be (as I dyd) to teache the vse of sub=traction the playner, therfore my fy=gures must stande thus when I haue ended my worke.

S. So I wolde thynke by that you taught me before, And nowe I thynke I coulde sub=tract any lyke sūmys.

```
  40308964
  20003428
  10002342
  10101461
  ———————
     111
  201733
```

M. So may you yf you haue marked, what I haue taught you. But bycause this thynge (as all other) must be lerned suerly by often practyse, I wyll expound here .ii. ex=amples to you, whiche yf you often do practyse, you shal be rype and per=fecte to subtracte any other summe lyghtly, for in it is contayned all the obseruances of hole nombre. And by=cause you shall perceaue sumwhat bothe howe to do it, and also whether it be well done, when you haue pro=ued to do it, therfore haue I written
vnder

Subtraction.

```
  308964         125614
  103145 ⎤        342 ⎤
  102597 ⎬        681 ⎬
  101024 ⎦        201 ⎦
     11            11
  ─────          ──────
   2198          124390
```

vnder them bothe theyr remayners. S. Syr I thanke you, but I thynke I myghte the better do it, yf you dyd shewe me the workynge of it. M. Yea but you muste proue your selfe to do some thynges that you were neuer taught, or els you shall not be able to do any more then you were taught, ŷ were rather to learne by rote (as they call it) thē by reasō. And agayne there is nothing in this eāple or any other of hole nōbre, but I haue taught you the rules of them al ready. S. Then I trust by practise to attayne ŷ vse of it. And is this al ŷ I shal learne of sub=traction? M. Yea, sauynge that (as you haue sene in Addition) there are broken nombres or fractions, in whi=che the workyng is not moch vnlyke,

yet

Subtraction.

yet wout some instructiōs be geuē of it, it myghte seme to a lerner moꝛe difficult thē in dede it is, therfoꝛe I wyl bꝛeuely shewe you ẏ vse of it onely by one exāple oꝛ .ii. A certayn man owed to me 14 li. 12 s̄. 8 d̄. of which he payd me at one time 4 li. 6 s̄. 8 d̄. at another tyme 3 li. ⸫ at another tyme 2 li. 3 s̄. 4 d̄. ⸫ laste of all 6 s̄. 8 d̄. now wold I knowe, what remayneth vnpayde yet: Therfoꝛe I set my sumes thus.

S. Syꝛ I pꝛaye you, why do you wꝛite 2 li? foꝛ the cōmon speache vsyth rather to saye 40 s̄. M. We muste here vse the denomynation that is grea-

	li.	s̄.	d̄.
	14	12	8
	4	6	8
	3		
	2	3	4
		6	8

teste in any sūme, so that we may not wꝛyte, accoꝛdyng as we vse to speake, sayng 16 d̄. 18 d̄. oꝛ lyke wayes : 7 grotes 8 grotes 24 s̄. 40 s̄ 48 s̄. ⸫ soche other, but we must wꝛyte euery denominatyon, that is in any summe, by it selfe, namely shyllynges and
poun=

Subtraction.

poundes so must we wryte for these summes nowe namyd, 1 s. 4 d. 1 s. 6 d. 2 s. 4 d. 2 s. 8 d. 1 li. 4 s. 2 li. 2 li. 8 s. and so forth of other lyke. S. So that we may not wryte in arythmetyke pennes, when the summe amountyth to shyllynges, nor shyllynges when the summe maketh poūdes Now yf it please you ende your exāple.

M. When my sūmes are so set as I shewyd then must I begynne with the smalleste denominatyon, saynge 8 4 8 are 20, whiche summe bycause it is pennes, and 12 pennes do make 1 s. I must take from that 20 (whiche commeth of the .iii. parcelles 12, and for them wryte 1 betwene the lynes, vnder ẏ shyllynges: then the 8 d. that remayneth, I take out of the hygheste summe (which is 8 also) and then remayneth naught
wher=

li.	s.	d.
14	12	8
4	6	8
3		
2	3	4
	6	8
	1	
4	16	

Subtraction.

wherfore vnder the pennes, I wryte nothynge. Then come I to the ſhyllynges, and gather y̆ parcelles 6, 3, 6, whiche with the 1 betwene the lynes make 16, that muſt I take out of the ſumme, that is ouer it : but ſeynge y̆ ſumme is but 12, I can not take 16 out of 12, I muſte borowe 1 of the 14 li. & put to the 12, & that maketh 32, for 1 li. is worth 20 s̈. then take I 16 out of 32, and there reſteth 16, to be writen vnder the ſhyllynges. Then come to the poundes, whoſe parcelles are 2, 3, 4, that is in all 9, and 1 more muſte I adde therto, bycauſe of the 1 that I borowed before vnto the 12 s̈. and then is there 10 which I muſt take out of 14, ſo doth there remayne 4 to be written vnder the poundes : ſo dothe my remayner appere to be 4 li. 16 s̈. S. This do I perceaue very well, and yf there be none other thyng to be learned in ſubtraction, then maye I come to multyplycation, for that you rekened next.

     G. M.

Subtraction.

M. we haue done n dede wt the art of Subtractiõ, as touchynge the workyng. But yet before we go to Multiplication, I wyll enstructe you howe to examyn your worke, whether it be well done or no, & that is by castynge awaye 9, as often as you can fynde it (as you dyd in Addytion) sauynge yt you must here exampne the hyghest nombre alone, and note the resydewe of it, at a lynes ende, as you dyd in addition. And when you haue done whith the hyghest nomber, then exampne all the other together, castynge thense 9 as often as you can : and yf the remayner be lyke the other, then haue you done. But yf you haue dyuers denominations in your summe, yet for them all shall you make but one seuerall lyne (as you dyd in Addition) remembrynge to begynne the examination at the greatest denominatiõ, and to double the remayner of poundes, and triple the remayner of shyllynges, as you dyd also in Addi=
tion

Subtraction.

tion. As for a profe, I wyll examyne this worke, wherin the hyghest lyne I fynd of poundes 14, frō thence I bate 9, and there resteth 5, whiche I do double (bycause they are poundes, and then are they 10, therto I

li.	s.	d.
14	12	8
4	6	8
3		
2	3	4
	6	8
4	16	

adde the 12, and it maketh 22, from which I take 9 twyse, and there resteth 4, which (bycause they are shyllynges) I triple, and then are they 12, therto I adde the 8, & then are they 20, thence take I twyse 9, and yet resteth 2, whiche I write at the one ende of a lyne, thus. 2 ——— Then I exampn all the other parcelles, and ẏ remayner together, euery denomination by it selfe. And fyrste of poundes I fynde 4, 3, 2, 4, that is 13, frome whiche I take 9 and there resteth 4, that do I double and it maketh 8, to it do I put the shyllynges 6, 3, 6, 16, that is 31, for the 1 betwene the ly=

G.ii.   nes

Subtraction.

nes must not be reckened, nor none in that space) and that maketh in all 39, where hence I take 9 .iiii. tymes, & there remayneth 3, that do I take .iii. tymes and it is 9, wherfore I caste it awaye: then do I take the pennes 8, 4, 8, that make 20, from whiche I take 9 twyse, & there resteth 2, which I write at the other ende of ye profe lyne, and bycause I se that those .ii. nombres are equall, I saye ye I haue well wrought. And yf you wyll, you maye make for euery denomination a lyne, as you lerned in addition: but then must you begyn your examination at the smalest denomination, as you dyd in Addition, for theyr profe is al together lyke, sauing that in Addition you exampned the nethermost summe alone, & al the other together, & in Subtractiō, you must exampne the highest nōber alone, & al the other together. And yf you marke it well it is euen all one, for ye summe that in Additiō is lowest, in subtractiō is hyghest, and

Subtraction.

and ẏ ſume is called ẏ groſſe oꝛ totall ſūme. Therfoꝛe yf you marke what I ſaid in Addition, you may eaſely perceaue, what is to be done foꝛ the pꝛofe of ſubtractiō, & to thentent ẏ you may perceaue it ẏ better, I wyl ſhewe you another pꝛofe of Subtractiō, & that ſhall be by Additiō, thus. Drawe vnder the loweſt nōber (which is your remayner) a lyne, then adde ẏ nomber, & all ẏ other, that you dyd ſubtracte befoꝛe together, & wꝛyte that that amōuteth vnder the loweſt lyne, & yf ẏ ſūme that cometh therof be equall to the hygheſt of the ſubtractiō, thē was ẏ ſubtractiō well wꝛought, els not, as

Groſſe or totall ſumm.

foꝛ exāple, in ẏ laſt ſūmes, which ſtode thus. Fyꝛſt I adde 8, 4, 8, that maketh 20, wherof I take 12 awaye (bycauſe they make one ſhyllyng) and wꝛyte foꝛ them 1 vnder the ſhyllin=

li.	s.	d.
14	12	8
4	6	8
3		
2	3	4
	6	8
1	1	
4	16	

G.iii.

Subtraction.

ges, and the 8 that is lefte, I write benethe the lowest lyne, then adde I y̆ shyllynges 6, 3, 6, 1, 16, that make 32, from whiche I take 20, and for it I write 1 vnder the poundes, and the 12 that remayneth, I write vnder the shyllynges. Then come I to the pou̅des, addynge them together, whiche are 4, 3, 2, 1, 4, y̆ maketh 14, the̅ do I write 14 vnder the li, & so haue I ended y̆ additio̅, & I se y̆ the lowest lyne of no̅ber & the hyghest be lyke, wherfor I know y̆ I haue wel done, for my figures appere thus.

	li.	s.	d.
And thus now haue	14	12	8
I taught you y̆ arte	4	6	8
of Subtractio̅, and	3		
the meanes to proue,	2	3	4
whether it be well		6	8
wrought or not.	1	1	
	4	16	
Now, and you reme̅ber, I omytted in tea	14	12	8

chyng the profe of Additio̅, one way, whiche I sayd was by Subtraction. S. Truth it is, and then was it deferred

Subtraction.

red, bycause that I had not then lear
ned the feate of subtraction, wherby
I shulde haue proued it, but now (I
thanke you) I haue well learned the
art of Subtractiō, and the proues of
it, both by 9, & by Addition, And now
I wold be glad to knowe how I may
proue Addition by Subtraction. M.
Then marke you this : whē you haue
ended your addition, take the nōbers
all that you dyd adde to the hygheste
summe, and deducte or subtracte thē
frō the grosse sūme y̌ doth resulte, & yf
the remayner be lyke to y̌ hyghest nō=
bre, then haue you done well, els not :
as for exāple. I take one of y̌ sūmes y̌
I dyd adde before, whiche was this.
Then do I come to the
mydle nōber, & subtract
y̌ frō y̌ nether nōber be=
gynnyng at y̌ left hand,
          106800
            4400
          116200

and fyrst I saye 0 out of 0, there re=
mayneth 0, that write I vnder an o=
ther lyne. Then agayne 0 (in the se=
conde place) from 0, remayneth 0,
        G.iiii.   vn=

Subtraction.

vnder it I write o alſo. Then in the thyrde place, 4 out of 2 wyll not be, therfore I adde to that 2, 10, & make it 12, from that I take 4, and there reſteth 8. Then ſay I farther 9 in ẙ iiii. place and 1 (which I muſte adde for the 10 borowed before) make 10, that muſt I take from 6, & bycauſe I can not, I adde, to the 6, 10, and then is it 16, from thence I take 10, and there reſteth 6 to be writtē vnder them. Then in the .v. place, where I fynde nothynge written, I muſt ſet 1 for the 10 laſt borowed, and that 1 do I take from the 1 vnder hym, and ſo remayneth naught, wherfore I write downe a cipher o. Now haue I done with the ſubtraction, and yet in the groſſe ſumme remayneth 1 whiche I muſt ſet ryght in the ſame place, in ẙ remayner, and ſo the re= mayner appereth to be lyke vnto the hygheſte ſumme of the addition, as here appereth.

```
106800
  9400
------
116200
106800
```

wherfore

Subtraction.

wherfore I saye, that the Addition was wel wrought, and thus may you do in any other summe of one denomination or many. Therfore now wyll I make an ende of Subtraction, and wyll instructe you in Multiplicatiõ.

℃ Multiplication.

Multiplication is an operation by .ii. summes producynge the thyrde, which so many tymes shall contayne the fyrste, as there are vnytes in the seconde, And it serueth in the stede of many addicions : as for example. Yf I wolde knowe how many are thyrtye tymes 48, yf I shuld adde 48, thyrty tymes it wolde be a longe worke : therfore was thys worke of Multiplication deuysed, which shall do that at ones, that Addytion shuld do at many tymes. S. I perceaue the cõmodyte of it partely, but I shall not see the full profyt of it, tyll I know y̨ hole vse of
G.v.   it.

Multiplication.

it. Therfore syr, I beseke you, teache me the workynge of it. M. So I iudge it best, but bicause that great summes can not be multiplied, but by the multiplycatiō of dygettes: therfore I thynke best, to shewe you fyrst the arte of multyplyenge them, as whē I say, 8 tymes 8, or 8 tymes 9. &c. And as for the small dygetes vnder 5 it were but foly to teache any rule, seynge they are so easy that euery chylde can do it: but for the multyplycation of the greatter dygetes thus shall you do. Fyrst set your dygetes one ouer the other ryght, then loke howe many eche of them lacketh of 10, and wryte that agaynst eche of them, and that is callyd the dyfferences: as yf I wolde knowe how many are 7 tymes 8, I muste wryte those digettes thus.

Multiplication of Digitys.

The Difference.

```
8
7
```

Then do I loke how moche 8 doth dyffer from 10, and I fynde it to be 2, that 2 do I wryte at the ryghte hande of 8, thus:

Then

Subtraction.

   8   2 Then do I take ẏ dyfference
   7 of 7, lykewayes from 10, ẏ
is 3, and I write that at the
ryght syde of 7, as you se in this ex=
ample.   8 2
Then do I drawe a lyne vn=
der them, as in additiō, thus.   7 3
Then do I multiply the two dyfferē=
ces, sayeng : 2 tymes 3 make 6, that
must I euer set vnder the differences
beneth the line : thē must I take ẏ one
of the differences (which I wyl for all
is lyke) from the other digette, not
from his owne, and that ẏ   8 2
is lefte, muste I write vn=
der the digettes, as in this   7 3
example.   5 6
Yf I take 2 from 7, or 3 from 8, there
remayneth 5, that 5 must I wryte vn=
der the dygettes, & then there appe=
ryth the multyplycatiō of 7 tymes 8,
to be 56. And so lyke of any other dy=
gettes, yf they be aboue 5, for yf they
be vnder 5, then wyll theyr dyfferen=
ces be greater than them selfe, so that
they

Multiplication.

they can not be taken out of them, & agayne suche lytle summes euerye chylde can multiplye, as to say, 2 tymes 3, or 4 tymes 5, & suche lyke. S. Truthe it is, and seynge, me semeth, that I vnderstand the multyplyenge of yͤ greatter dygettes, I wyll proue by an example how I can do it: I wold knowe howe many are 9 tymes 6. M. It is al one in valewe to say 9 tymes 6, or 6 tymes 9, but yet the order is beste to put the lesser summe fyrste, sayenge, 6 tymes 9, & so of all other summes. S. then wolde I know how many are 6 tymes 9, therfore I set the dygettes, thus.

        9
Then do I set ther dyfferences at theyr ryghte syde, the   6
dyfference of 9, which is one agaynst it, & the difference of 6 whiche is 4 agaynst it also, as in this example.

  9 1 And vnder them I drawe a
  6 4 lyne, then do I multyply the
  ――― dygettes together, sayenge 1
tyme 4 maketh 4, that 4 do I wryte
           vnder

Multiplication.

vnder the dyfferences thus.
Then take I one of the dy=
fferences from the other dy=
get, as 1 from 6, oz els 4 frō
9, and eche wayes there resteth 5, whi=
che I do wryte vnder the dygettes,
and so apperyth the multiplicatyō of
6 tymes 9 to be 54. Thus I see the
feate of thys maner of multyplycatiō
of dygettes. M. Nowe mought you go
strayght to the multiplycation of
greatter nōbers, saue ÿ both for your
ease & suertye in workynge, I wyll
drawe you here a table, wherby shall
appere the multiplycation of all dy=
gettes, & this is it.

```
9   1
6   4
─────
5   4
```

℧ A table to multiply all
dygettes by.

Multiplication.

1	1	2	3	4	5	6	7	8	9
	2	4	6	8	10	12	14	16	18
		3	9	12	15	18	21	24	27
			4	16	20	24	28	32	36
				5	25	30	35	40	45
					6	36	42	48	54
						7	49	56	63
							8	64	72
								9	81

In whiche figure when you wolde knowe any multiplicatiō of digettes, seke your fyrste oꝛ laste digette, in the blacke squares, and from it go ryght forth towarde the ryghte hande, tyll you come vnder the figure of your se=conde diget whiche is in the hyghest rowe, and then the nomber that is in the metynge of theyꝛ bothe squares, is the multiplycation that amoūteth of them. As yf you wolde knowe by this table the multiplicatiō of 7 ty=mes 9, seke fyrst 7 in the blacke squa=res, and then go ryghte forth toward
the

Multiplication.

the ryght hande, tyll you come vnder 9 of the hyghest rowe, and in the me=tynge of theyr squares, you may se 63 whiche is theyr multiplycation. S. Thys is very good and reddy, and so may I kynde the multyplycation of any digettes. But nowe how shall I do in greater sumes? M. When you wolde multiply any summe by an o=ther, you shall marke, that it is the metest order to set the greatest nom=ber hyghest, which is the place of the nomber that must be multyplyed : & lyke wayes the lesser nōber vnder it for that is the place of the multiplyer or multyplycatour, that is to saye the nomber, by whyche multyplycation is made, and is in Englyshe all wayes put before this worde, tymes, in soch speakynge, when I say, 20 tymes 70, and the nomber that folowyth this worde, tymes, is that whiche must be multyplyed. Therfore when I wolde multiplye one nomber by an other, I must write the greatest hyghest, and the

To Multiply greater sume.

Multiplyer.

Tymes.

Multiplication.

the leſſer vnder it, as in additiō. And vnder them muſte I drawe a lyne, as for exāple. Yf I wold multiply 2 6 4, by 2 9, I muſt ſet them thus. Then muſt I multiply euery figure of the hygher rowe, by euery fygure of the nether rowe, and that that amountyth I muſt ſette vnder the lyne, as thus: fyrſt I do multyplye 4 by 9, ſayenge 9 tymes 4 (or 4 tymes 9 which is all one) and that makyth 36, as the table before of dygettes doth declare, of that 36 I muſt wryte the 6 that is the dyget vnder the 9, and the 3 in the nexte place towarde the lefte hande.

```
  2 6 4
    2 9
  —————
```

Then come I to the ſeconde figure of the hygher rowe & ſay, 9 tymes 6, make 54, of which I write the 4 vnder the 3, and the 5 vnder the next place (as reaſon wylleth me) thus.

```
  2 6 4
    2 9
  —————
    3 6
```

Then come I to the nexte fygure which is 2, and do multiply it by 9, & that maketh

```
  2 6 4
    2 9
  —————
  5 3 6
      4
```

Multiplication.

18, wherof I write 8 vnder the thyrd place, and the article 1, in the fourth place, thus.
And then haue I ended the fyrst figure of the multy=plyer. Then begyn I with the nexte figure, and multi plye it into all the hygher fygures, as thus.
Fyrst 2 tymes 4, make 8, that do I write vnder the se= conde place (for euer more ẏ diget, or fyrst figure of ẏ mul tiplication ẏ amoūteth of the fyrst fi= gure of the higher nōber, must be set vnder the multiplyer of it, & the other in theyr ordre toward the lefte hand.
S. I vnderstande you thus, that the digette of the summe amountynge of the multiplycation of the fyrst figure of the hygher rowe, by the fyrst figure of the lower rowe or multiplyer, must be set vnder the fyrste place : and that amounteth of the same fyrste figure, by the seconde multiplyer, must be set

```
  2 6 4
    2 9
  -----
  1 5 3 6
    8 4
```

```
  2 6 4
    2 9
  -----
  1 5 3 6
    8 4
    8
```

H   under

Multiplication.

vnder the seconde place: and so of the other, yf there be moze multyplyers. M. So meane I in dede: and yf there amount but a dygete, then must it be sette vnder the same. And now to go forth, I multiplye by the same 2, the seconde fygure of the hygher rowe, which is 6, sayeng 2 tymes 6, make 12, where of I write the dyget 2 vn= der the thyrde place, and the artycle 1 I write vnder the fourth place.

Then do I multiplye the laste fygure of the hygher summe, by that same 2, sayenge, 2 tymes 2 is 4, whiche I write vnder the fourth place, and so haue

```
  2 6 4
    2 9
  -----
  1 5 3 6
  1 8 4
    2 8
```

I endyd the hole multiplycation, and the summes stande thus.

Than must I drawe a lyne vnder all those summes that amount of multiplycatiō, & must adde all them into one summe, as in this example

```
  2 6 4
    2 9
  -----
  1 5 3 6
  1 8 4
  4 2 8
```

you maye see, where in the fyrst place

I

Multiplication.

I kynde but 6, and ther=
fore write I it vnder the
lyne: then in the seconde
place 8, 4, 3, make 15 wher
of I write 5, and kepe 1 in
my mynde, and so forth,
as you lerned in Additiō

```
  2 6 4
    2 9
  -----
  1 5 3 6
  1 8 4
  4 2 8
  -------
  7 6 5 6
```

and so apperyth the hole summe to be 7656, which amountyth of the mul= typlycation of 264 by 29. S. If ther be no more to be obseruyd in it, then can I do it, I suppose, as by this ex= ample I shal proue: I wold multiply 1365 by 236, wherfore I set thē thus.

Then do I multiply 5 by
6, sayeng 6 tymes 5 make
30, of whiche I write the

```
  1 3 6 5
    2 3 6
```

cypher in the fyrst place, & the article
3 in the seconde place.
Then do I by the same 6
multiplye the seconde fy=
gure of the higher sūme,

```
  1 3 6 5
    2 3 6
  -------
      3 0
```

which is 6, sayenge, 6 tymes 6 make 36, of which I wryte the 6 vnder the seconde place, and the 3 vnder the

H.ii. thyrde

Multiplication.

thyrde place.
Then do I multiply the
thyrde figure which is 3
by the same 6, & that ma=
keth 18, of that I set ye 8
vnder the thyrd place, and the 1 in the
fourthe place.
Then come I to the laste
fygure of ye hygher sume,
and multiply it by 6, say=
enge, 6 tymes 1 make 6,
that do I write vnder ye fourth place.

And so haue I ended the fyrst
multyplyer: then begynne I
with the seconde multiplyer,
and say fyrst, 3 tymes 5 that
maketh 15, of which I set the
5 vnder ye seconde place (bycause that
the multiplyer is there) and the 1 I
set vnder the thyrde place.
Then come I to the secōd
figure that is 6, and mul
typlye it by 3, whiche ma=
keth 18, of which I set ye
8 vnder the thyrde place,

```
1 3 6 5
  2 3 6
-------
  3 3 0
      6
```

```
1 3 6 5
  2 3 6
-------
1 3 3 0
    8 6
```

```
1 3 6 5
  2 3 6
-------
1 3 3 0
  6 8 6
```

```
1 3 6 5
  2 3 6
-------
1 3 3 0
  6 8 6
    1 5
```

and

Multiplication.

and the article 1, in the fourth place.

```
  1 3 6 5
    2 3 6
  -------
  1 3 3 0
  6 8 6
  1 1 5
      8
```

Then come I to the thyrde figure which is 3, and multiply it by 3, sayeng, 3 tymes 3, make 9, which bycause it is but one dygette, I set vnder the fourth place.

And then cōmynge to the last figure 1, I multiply it by 3, and it maketh 3, which I set in the fifte place, & then haue I ended .ii. of the multiplyers, and the summes stande thus.

```
  1 3 6 5
    2 3 6
  -------
  1 3 3 0
  6 8 6
  1 1 5
      9 8
```

```
  1 3 6 5
    2 3 6
  -------
  1 3 3 0
  6 8 6
  1 1 5
    3 9 8
```

Then come I to the thyrde multiplyer, and multiply it into euery figure of the hygher summe, and fyrst I saye 2 tymes 5 make 10, of which I set the cypher vnder ẏ multyplyer, in the thyrde place, and the article 1, in the fourth place. And so multyplyenge the seconde fygure 6, by that same 2, there amoūteth 12, wherof I write the dyget 2, vnder the fourth place, and the article

```
    1 3 6 5
      2 3 6
    -------
    1 3 3 0
    6 8 4
    1 1 5
  3 9 8
  1 0
```

H.iii.

Multiplication.

cle 1 vnder the fyrſte place.
Now do I multiply by the
ſame 2, the thyrde figure
of the hygher ſume, whiche
is 3, and that maketh 6,
whiche I ſet vnder ẏ fyrſte
place. Then come I to the
laſte place, and multyplye
that 1 by 2, and there a=
mounteth 2, whiche I ſet in the ſyrt
place, & thē doth ẏ ſūmes ſtand thus.
And ſo haue I ended the
hole multiplycatiō. But
now (as you taught me)
to knowe what this hole
ſume is, I muſt adde all
thoſe parcels together,
and then vnder the lyne
wyll appere the groſſe or
total ſume, ẏ is 322140.
M. That is well done. S. Then me
thynketh I wolde call it well done,
when I knewe whether I had well
done or no. M. It may be tryed by 9
as Addition was, but the ſureſt proſe
is

```
    1 3 6 5
      2 3 6
    1 3 3 0
      6 8 6
      1 1 5
      3 9 8
      1 1 0
        6 2
```

```
    1 3 6 5
      2 3 6
    1 3 3 0
      6 8 6
      1 1 5
      3 9 8
      1 1 0
          2
```

```
    1 3 6 5
      2 3 6
    1 3 3 0
      6 8 6
      1 1 5
      3 9 8
      1 1 0
        2 6 2
```

Multiplication.

is by Dyuyſion, and therfoꝛe I wyll reſerue that, tyll you haue learned tharte of Diuyſion. And before we paſſe from Multiplicatiō, I wyll yet ſhewe you other wayes of it, whiche are counted of ſome men both moꝛe reddye, & moꝛe certayne, of which the one dyfferyth nothynge from this ẏ I haue taught you, ſayenge that it doth vnderſtade all wayes the artycles and ioyne them to the next ſūme: & therfoꝛe I wyll declare it onely by and exāple: Yf I wold multiply 1542 by 365, I muſt ſet them as I ſayde befoꝛe, and then do I multiply 2 by 5 and it maketh 10, of whiche I wꝛyte the article vnder the fyꝛſte place, and kepe the dygitte 1 in my mynde.

```
1542
 365
————
   0
```

Then ſay I foꝛth, 5 tymes 4 do make 20, and ẏ one in my mynde are 21, therof I wꝛite the 1 vnder the ſeconde place, & kepe the 2 in my mynde. Then come I to the thyꝛde fygure 5, ſayeng, 5 tymes 5,

```
1542
 365
————
  10
```

H.iiii.   make

Multiplication.

make 25, and the 2 in my mynde, make 27, wherof I write the 7 vnder the thyrde place, and kepe the article 2 in my mynde. Then commynge to the last fygure I say 5 tymes 1 make 5, and 2 in my mynd make 7, that do I wryte vnder the fourthe place. And then haue I ended my fyrste multyplyer. Then do I lyke ways with the secōd multiplyer, sayenge, 6 tymes 2 make 12, therof I write the dygete 2 vnder the seconde place, and kepe the article 1 in my mynde. Then saye I forth, 6 tymes 4 maketh 24, and the 1 in my mynde make 25, so I set that 5 vnder the thyrde place, and kepe the 2 in my mynde. Then multiply I forth sayenge, 6 tymes 5 makyth 30, & 2 in my mynd make 32, wher of I write the 2 vnder the fourth

```
     1 5 4 2
       3 6 5
     ─────────
     7 1 0

     1 5 4 2
       3 6 5
     ─────────
   7 7 1 0

     1 5 4 2
       3 6 5
     ─────────
   7 7 1 0
         2

     1 5 4 2
       3 6 5
     ─────────
   7 7 1 0
       5 2
```

Mulitplication.

fourth place, & kepe the 3 in my mind.

```
  1 5 4 2
      3 6 5
  ─────────
  7 7 1 0
      2 5 2
```

Then do I multyplye the laste fygure 1 by 6, and it maketh 6, to that I adde the 3 in my mynde, and it maketh 9, whiche I write in the fyrst place.

And so haue I ended .ii. fygures of y͘ multiplyer, Thē with the thirde and last multyplyer do I lyke wayes, & saye fyrst, 3 tymes 2 make 6, whiche I write in the thyrde place vnder the multiplyer.

```
  1 5 4 2
      3 6 5
  ─────────
  7 7 1 0
  9 2 5 2
```

Then by that 3 do I multiply lykewayes the second figure 4, and it maketh 12 wherof I write the digette 2 vnder the fourth place, and the article 1 I kepe in my mynd.

```
  1 5 4 2
      3 6 5
  ─────────
  7 7 1 0
  9 2 5 2
        6
```

Then come I to the thyrd figure 5, sayeng : 3 tymes 5, maketh 15, and the 1 in my mynde make 16, therof I write the 6 vnder y͘ fyfte place,

```
    1 5 4 2
        3 6 5
    ─────────
    7 7 1 0
    9 2 5 2
        2 6
```

H.v.

Multiplication.

place, and kepe the artycle 1 in my mynde.
Then come I to the laste figure, whiche is 1, and multiplye it by 3, and it maketh 3, therto I adde the 1 in my mynde, and it maketh 4, whiche I write in the sixte place. And then haue I ended the multiplicatiō, & the figures stande in ordre, thus:

```
  1 5 4 2
    3 6 5
  ───────
  7 7 1 0
  9 2 5 2
  6 2 6
```

```
  1 5 4 2
    3 6 5
  ───────
  7 7 1 0
  9 2 5 2
4 6 2 6
```

whiche parcelles yf I adde into one summe, it wyll be 562830, whiche is ye grosse or total summe of all that multiplication. S. Well, this maner of multiplication I perceaue: but what other sortes haue you? M. There is one way that is wrought by a checker table, made thus.
And loke how many places your summe hath, that you wold multiplye, so many squares must you make in your table from

To Multplye by a checker table.

Mulitiplication.

from the right ſyde to the lefte, and ſo many frome the hygher parte to the lower, as there be places in your multiplyer. Thē ſet downe your greateſt ſumme fyrſt on the toppe of the table, euery figure (in dewe ordre) ī a ſquare alone, I meane in thoſe ſquares, that be open and vncroſſed. And lyke⸗ wayes in thoſe lyke ſquares, at the ryght hande ſet downe your multiplicator, the laſte figure in the hygheſt place, & ſo downwarde, that the fyrſte figure maye be in the loweſt place. S. Syr, yf it pleaſe you, me thinketh thē I vnderſtande you beſte, when you do not ſtande longe in tellyng ẏ rule before examples, but propoſe ſome example, & then in declaryng it, brynge in the rules with all. M. In dede that waye is eaſyeſt for a yonge learner, therfore wyl I euen ſo do. Take this example now : I wold multiply 2036 by 23. Fyrſt I conſyder that my grea⸗ teſt nomber hath .iiii. fygures or pla⸗ ces, and therfore I make ſo many

roumes

Multiplication.

roumes betwene lynes, thus.
Then I see that of my mul=
typlyers there are .ii. wher=
fore I drawe so many lynes
a crosse the other, that there
may be 2 roumes betwene thē, thus.
But you muste not for=
get to let the endes of the
lynes runne out (as it ap=
pereth in this patrone) for
in those open squares must your .ii.
fyrst nombers, and all the totall sūme
be set, Then drawe a crosse barre tho=
rough euery close square, so that it
may reche down to the lowest ouerth
warte lyne, as in this fourme.
And thē is your chec=
ker fourme prepared
then set downe your
fyrste or greateste sūme on the toppe,
and your multiplier on y ryght syde
in the open squares, thus.

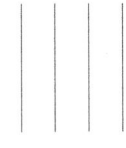

Then begynne to
multiply the first
fygure of the hy=
gher

Multiplication.

gher summe by the hygheste of the multiplier, sayenge: 2 tymes 6 maketh 12, that 12 must you write in that square that is agaynst the 2 and the 6, but of suche maner that the dygitte be set in the nether corner of y^e square, and the article in the hygher corner, as you may se in this example.

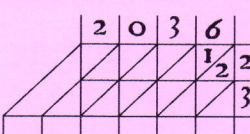

And so of every other multiplication, what euer amoūteth you must write in the commen square, whiche is agaynste bothe those fygures, by which you do multiplye, and yf that sume do make but one dygette, then must it be set in the lower corner of the square, but yf it make an article, then write the article in the hygher corner, and lette the cypher go (yf you wyll euermore) for here it serueth for nothynge, seynge the lynes do dystinct the places, but yf the summe amountynge of soche multyplication, do make a myxte nōber, then wryte ye article in the higher corner

Multiplication.

corner, and the dygete in the lower corner, as I dyd by that 12, then whē you haue multiplied and ended the fyrſt fygure, come to the nexte, and multiplye it in lyke maner: as in ſay‍enge, 2 tymes 3 is 6, that 6 bycauſe it is but a dygete, you ſhall ſet in the nether corner of the ſquare, nexte vn‍der 3, as here foloweth.

Then go forth ſay‍enge, 2 tymes 0 is 0, ſet that vnder the barre (yf you lyſt) in the thyrde ſquare.

Then forth, & ſaye, 2 ty‍mes 2 make 4, that ſette in the laſt ſquare vnder the barre, ſo haue you ended the fyrſt multiplyer: come therfore to ẏ ſecond and ſay, 3 tymes 6 make 18, of whiche ſumme the article 1 muſt be ſet aboue the barre, in that ſquare that is next to the 3 (as you ſe here) and the the 8 vn=

Multiplication.

```
 2 | 0 | 3 | 6
 4 | 0 | 6 |¹2 ²
           |¹8 ³
```
vnder the barre. Then saye, 3 tymes 3 maketh 9, sette it in the nexte square beneth the barre, then 3 tymes 0 is 0, write it in y̆ next square or let it go, for al is one. S. I perceaue, for here the lynes dy= stinct the placys wherfore ciphers do only serue, ⁊ therfore here they nede not to be. M. then say farther, 3 tymes 2 make 6, write that in the last square then wyll the hole fygure stand thus as here foloweth

```
          2 | 0 | 3 | 6
          4 | 0 | 6 |²2 ²
          6 | 0 | 9 |¹8 ³
```

S. Nowe coulde I (me semeth) do lyke agayne: but how shall I do now to gether the sũme? M. Marke fyrst the ordre of the places in this fygure, and so shall you perceaue the reason of getherynge them into a sũme. The slope barres do parte the places, so that the fyrst place is the loweste cor= ner (in all suche fygures) of the ne= thermost square, next the ryght hand and all the halfe squares betwene y̆

barre

Multiplication.

barre, and the nexte standeth for the seconde place, and so the roume be=twene that and the nexte barre is the thyrd place, and so forth. Now yf you perceaue this, then must you adde all the fygures of one place to gether, as yf you had an addityon of dyuers su̅=mes. S. Yf I vnderstande you ryght, then must I take here in this exa̅ple 8 to be in the fyrst place, 9, 1, & 2, in the seconde. 0, 6, 1, in ẏ thyrde. 6, 0, in the fourthe, 4 in the fyfte, and the syxte place hath no figure. M. you say wel, & the reason is bycause the multipli=cation seruynge to that square, made but a dygette. S. Then it is all one, as if they stode thus,
M. euẏn so it is, and nowe adde this summe, & there wyll appere the totall of ẏ multiplycation to be 46828. And yf you wyll se the agremente of this ma ner of multiplication, and the other that you lernyd before: then multiply those two summys (that is 2036 by 23)

$$\begin{array}{r}12\\061\\46098\\\hline\end{array}$$

Multiplication.

23) after ẏ fyrſt maner without ſqua=
res. S. You meane to ſet them thus
in order.   2036
And then multiply 3 into 6,   2 3
make 18, 3 tymes 3 make 9,   ─────
3 times 0 is 0, then 3 tymes 2 make 6,
which muſt be ſet thus.   2036
Then do I lyke ways with ẏ   2 3
ſeconde multiplyer, ſayenge,   1 8
2 tymes 6 make. 12, 2 tymes   609
3 make 6, 2 tymes 0 is 0, and
2 tymes 2 make 4, whiche when I
adde to the other, then wyll the hole
multiplycation ſtande thus.   2036
M. So that you may ſe in   2 3
euery place the ſame figures   1 8
here, as they were in the mul   609
typlycation by ſquares,   12
though they differ in heyght   406
& lownes of places, but being addyd
together they make one ſumme. And
thus now ye haue lerned .iii. ſortes of
multiplicatiō, which you liketh beſt ẏ
may you vſe. Yet are there other for=
mes, but ſyth they nothing differ frō
                           I.   theſe

Multiplication.

these thre in effecte, but onely in set=
tynge of the nōbers, J wyl ouerpasse
thē tyll a moꝛe meter place and tyme,
And nowe wyll J enſtruct you in Di=
uiſion, ſo that you thynke your ſelfe
ſufficntly to perceaue, what J haue
taught you. S. Yes ſyꝛ J thanke you,
but J do not ƿceaue how to erampne
my woꝛke, to trye whether J haue
well done oꝛ no. M. That is cōmenly
vſed by the pꝛofe of 9, as you learned
befoꝛe in Addition, and Subtractiō,
ſaue that it hath this wayes dyuers
from them. Fyꝛſt you muſte make a
croſſe after this maner.

Then muſt you exampn your
ſumme that ſhulde be multy=
plyed, and loke what remayneth after
caſtynge awaye of 9, that ſet you at
the one ſyde of the croſſe, then exa=
mpne the multyplyer, & what ſo euer
remayneth in it after caſtynge awaye
9, as often as you can, wꝛite that at
the other ſyde of the croſſe : then muſt
you multyply thoſe .ii. nombꝛes toge=
ther

Multiplication.

ther, and loke what amouteth therof, yf it be vnder 9, wryte it at the hygher parte of the croſſe : but yf if be aboue 9, then take thence 9, as often as you can, and wryte the reſte at the head of the croſſe, as in ẏ laſt example of mul tiplication : the nomber to be multy= plyed is 2036, wherin is ones 9, and 2 remayneth, whiche I wryte at one ſyde of a croſſe, thus.

Then do I exampne the mul= typlyer, whiche is 23, wherin there is no 9, but 5 in all, ẏ 5 ther= fore I ſette at the other ſyde of the croſſe, thus.

Then do I multyply 5 by 2, and it maketh 10, from which I withdrawe 9, and there reſteth 1, that 1 do I ſet at ẏ head of the croſſe, then do I exampne the groſſe ſumme amountynge of the multiplycation, whiche is 46828 wherin I fynde 9 thre tymes, and 1 remaynynge, that 1 I ſet at the fote of the croſſe : and then I ſe it to agree with the other 1

H.ii.   at

Multiplication.

at the toppe of the crosse, and so know I that I haue done well: for yf they two dyd dyffer, then were my worke vayne, and the multiplication false. This is the comen profe, but the most certayne profe is by Diuisiō, of whiche I wyll anone enstructe you. S. Syr, what is the chyefe vse of Multiplication? M. The vse of it is greater then you can yet vnderstand, how be it these playne commodities it hath, that yf you wolde resolue any greate and hole valeure into many small & lesse portiōs: as yf you wolde chaūge poundes into shyllynges, pennes or any other greater or smaller parcels, by Multiplycatiō you shall do it spedely & easely. Also yf you shulde nede to adde one summe to it selfe, or any other oftē tymes, you shall do it moch more spedely, reddely, easely, and surely, then by often and sundry Additiōs: Take you these cōmodities grosely shewed for an answere at this tyme, & hereafter I wyll more abūdant=
ly

Multiplication.

ly make you to perceaue ẏ vſe of it. S.
Well syr, then in Diuiſiō I pray you
to enſtruct me. But me thynketh by ẏ
name of it ẏ it ſhold be al one w̄ Mul
tiplycatyō : for I call that Dyuiſion,
when any thynge is parted into dy=
uers and many partes. M. You take
it, as it is taken commenly, how be it
yf you marke well you ſhall perceaue
that it is quyte cōtrary to Mutiply=
cation, and doth not part one thynge
or fewe thynges into many, but con=
trary wayes it bryngeth many par=
cels into fewe, but yet ſo, that theſe
fewe takyn together are equal in va
lure to ẏ other many, for by Diuyſiō,
pennes are turned into ſhyllynges,
& ſhyllyngs into poūdes : as for exam
ple, of 120. s̄. it maketh 6. li. ſo are
120 turned into 6 which is a ſmaller
nomber, but then yf you conſydre the
denomynotours, you ſhall ſe that
they are ſoche, that one of the later is
equall to 20 of the fyrſt, and ſo in val=
lewe the ſummes are one, though in

     I.iii. nom=

Multiplication.

nomber they do farre dyffer, and the later summe is the lesser, and so is it al wayes in Diuisyō, how be it yet in the workyng, the summe is parted by an other, & therof doth it take y̌ name S. I thynke I shall better vnderstād the reason of the name, whē I knowe the vse of the worke, therfore nowe wolde I gladly learne that.   Ma.

¶ Diuision.

Iuisiō is a ptitiō of a greater sūme by a lesser, therfore as you may perceaue, vnto Diuision are requyred .ii. nōbers, the fyrst, which shuld be diui= ded, & that must be y̌ greater, and the second, by which y̌ other must be diui= ded, & that is the lesser, & is called the diuisor. The fyrst must fyrst be writtē, & the secōd so set vnder it, y̌ the last fi= gure of y̌ lower nōber be right vnder the last of y̌ hygher, cōtrary wayes to the worke of y̌ other kindes of Arith= metyke, for in them the two fyrst fygu res were set euer mete one vnder the
other

Diuision.

other but in Diuision ye last fygures must be set mete, except it chaunce so that the laste fygure of the Dyuysor be greater then the last of the hygher nomber, for then you shall set the last of the Dyuysour, vnder the last saue one of ye hygher nōber, as for exāple, yf you shuld diuide 365 (which are ye sūme of ye dayes of a yere) by 28 (whi= che are the daies of a cōmen moneth) then shulde you set them thus.   365
But yf you wold diuide those   28
365 dayes, by 52 (which is ye nōber of wekes in one yere, then shuld you set them thus. Lyke wayes yf I   365
wold diuyde the same 365, by   52
4, which is the sūme of ye quarters of a yere, then must I set thē thus.   365
S. Syr this do I vnderstand,   4
but how nowe shulde I do to diuide the one by the other? M. You muste begynne with the laste figure nexte ye lefte hande, and se how many tymes the laste figure of the Diuisor may be taken out of ye laste figure of the ouer

H.iiii.   nom=

Diuision.

nomber, that shall you note within a croked lyne toward your right hand, as for example: I wold diuide 365 by 28, then set I those two sūmes, thus.

  965
  28

And I loke how many tymes I may fynde 2 (whiche is the last figure of the diuisor) in 3, (whiche is the last of the nōber to be diuided) and consideryng that I can take 2 out of 3 but ones, I make a croked lyne at the ryght hande of the nombres, and within it I sette 1, and

Quotient Nombre.

that is called the quotient nombre: then bycause that whē 2 is taken out of 3, there remayneth 1, I must write that one ouer 3, and deface or cancel the 3 and the 2, then wyll the figures stande thus.

Then must I go to the nexte figure of the di=uisor, and take it lyke

   1
   3̸65 (1
   2̸8

wayes so many tymes out of the fi=gures that be ouer it, and loke what doth remayne, that must I write ouer them, and cancell them, as in this ex=
        ample

Diuision.

ample : J take ones 8 out of 16, and there remayneth 8, whiche J muste set ouer the 6, & cancell or crosse out the 16 and the 8 of the diuisour. And then wyl the figures stand thus. And so haue J ones wrought S. So J perceaue that you take ẏ nether figure not onely out of ẏ other

18
3̷6̷5 (1
2̷8̷

that is ryght ouer hym, but out of ẏ with ẏ other also that remayneth before, and are wꝛiten towarde the lefte hande. M. So must you do, for you must so take the diuisor out of the ouer nomber, that there remayne not ouer it so great a summe as it is : for then were your worke in vayne. But yet agayn here must you marke, that when you seke howe many tymes the laste figure of the diuisour maye be founde in the nomber ouer hym, that you loke also whether you maye as often fynde all the figures folowyng in those that are aboue them, yf not, take your quotiente lesse by one, and

J.v. then

Diuision.

then proue agayne, and so styll, tyll you fynde a mete quotiēt. When you haue thus wrought ones, then must you begynne agayne, and write your dyuysor a newe nerer towarde the ryght hande by one place, as in this example, you shall set 2 vnder 8 and 8 vnder 5 thus.

Then as before seke how many tymes you may take the last dy=

```
   / 8
 3 6 5   ( 1
 2 8 8
     2
```

uysor out of the nombre ouer hym. S. That maye I do here 4 tymes. M. Truthe it is, that you maye fynde 2 foure tymes in 8, but then marke whether you can fynde the fygure fo=lowyng so manye tymes in the other that is ouer hym. Can you fynd 8 iiii. tymes in 5? S. No nother yet ons. M. Therfore take 2 out of 8 ones lesse. S. That is 3 tymes. M. Well then 3 tymes 2 make 6 : yf I take 6 out of 8 there remayneth 2, which 2 with the 5 folowyng make 25 in whiche summe I maye fynde 8 thre tymes also, and
        therfore

Diuision.

therfoꝛe J take 3 as a true quotyent, and wꝛite it within the crokyd lyne of the quotyent befoꝛe the 1 thus. Then say J 3 tymes 2 make 6, then 6 out of 8 resteth 2, therfoꝛe J cancell the 2, and the 8 and wꝛite ouer it the 2 that doth re= mayne thus.

```
 1̷8
 3̷6̷5    (13
 2̷8̷8
  2
```

Then do J take 8 as many tymes out of 25 sayeng 3 tymes 8 make 24, & yf J take 24 out of 25 there remayneth 1, so then J can cell 25 & 8, and ouer the 5 J set 1 thus.

```
  2
 1̷8̷1
 3̷6̷5   (13
 2̷8̷8
  2
```

And now haue J done with dyuidynge: foꝛ J can kynde my dyuysoꝛ 28 no moꝛe in the ouer summe. S. No, excepte you wolde parte the 1 that remayneth into 28 partes. M. That is well sayd, and so must we do in suche cases whē there remayneth any thynge, but J wyll lette that passe nowe, and wyll make

Diuision.

make you parfecte in hole Diuisyon, and wyll here after teache you peculyarly of broken nombre callyd fractions. And now yf you do perceaue the order of dyuisyon, then do you dyuide this summe 136280 by 452. S. Fyrste I set downe the nomber that shulde by dyuided, then do I set the dyuisor vnder it, so that the laste fygure of it, be ryght vnder the laste fygure of the ouer nomber. Then wyll it be thus. M.   136280
Can you take the last of   452
your dyuisor (whiche is 4) out of 1 which is the last of the ouer nomber? S. I had forgotten, bycause the laste of the dyuisor can not be taken out of the last of the ouer nōber, in as moche as it is the greater, therfore must I sette the dyuisor one place more forward towarde the ryght hande, thus. And then muste I loke   136280
how often I may fynde   452
the last fygure of the dyuisor (that is 4) in 13, whiche thynge I may do 3 tymes,

Diuision.

mes therfore do I say, 3 tymes 4 is 12, which I take out of 13 & there remayneth 1, then do I make at the ryght hand of my sūmes a croked lyne, and write before it my quotyente 3, and I cancell 13 and 4, and ouer the 3 I sette the 1 that remayneth, and then the figures stande thus.

Then do I multyply the same quotient into euery figure of the
$$\begin{array}{r} 1 \\ 1\cancel{3}6280 \ \ (3 \\ \cancel{45}2 \end{array}$$

diuisor, and withdrawe the summe y amounteth, out of the nombres ouer them: as fyrste I say, 3 tymes 5 make 15, which I take from 16, and there resteth 1, I cancell therfore 16 and 5, & write ouer the 6 that 1 that remayneth, thus.

Then do I say lyke wayes, 3 tymes 2
$$\begin{array}{r} 1\,1 \\ 1\cancel{3}\cancel{6}280 \ \ (3 \\ \cancel{4}\cancel{5}2 \end{array}$$

make 6, whiche I take out of 12, and there resteth 6, therfore I cancell the 12 and the 2, and ouer the 2 I write 6 that remayneth, thus.
$$\begin{array}{r} 1\,1\,6 \\ 1\cancel{3}\cancel{5}\cancel{2}80 \ \ (3 \\ \cancel{4}\cancel{5}\cancel{2} \end{array}$$

Then

Diuision.

Then shulde I set forward the diui=
sor into the nexte place towarde the
ryght hande thus.   1̸1̸6
M. But you may se ẏ   1̸3̸6̸280   (3
ouer ẏ 4 is no figure, 45̸2̸
therfore must you set   452
the diuisor yet forwarder by an other
place. And marke whē so euer it chaū
cheth so, that you shulde set forward
the diuisor, and that it can not stande
there, bycause there is no nōber ouer
the last place, or yf there by any, it is
lesser then the last figure of the diui=
sor, then must you remoue the diuisor
yet ones agayne: and bycause ẏ his
fyrste place of remouyng fayled hym,
therfore shall you write in the quotiēt
a cyphre o, & yf you shulde by chaūce
nede to do so often tymes, for euery
tyme write a cyphre in the quotient,
The reason of this wyll I showe you
hereafter. S. Then   1̸1̸6
must I set my sum   1̸3̸6̸280 (30
mes thus.   45̸2̸52
And bycause I re=   4

moued

Diuision.

moued the diuisor, so þ̄ I ouerskyp=
ped one place, I must wryte a cyphre
in the quotient: and then must I seke
a newe quotient, as in this example:
How many tymes 4 is there in 6, and
syth it can be but ones, therfore do I
wryte 1 in the quotient, & then say I, 1
tymes 4 taken out of 6 remayneth 2,
I cancell the 6 and the 4, and wryte
2 ouer thē. Then say I agayne, ones
5 out of 28 remayneth 23, I let the 2
stande as it dyd, and ouer the 8 I set
3, cācellyng þ̄ 8 & the 5 vnder it thus.

M. You myght as
wel haue sayd ones
5 out of 8 and so re=
mayneth 3: but now
go forth. S. Then
ones 2 out of 0, can not be, what shal
I nowe do? M. Borowe of the nexte
nomber that is behynde (for there is
230) and do as you learned in Sub=
traction in a lyke case. S. Then must
I borowe 1 of the 3 cōmynge behynd
nexte, and make that 0 to be 10, and
then

```
          2
        1̸1̸6̸3
       1̸3̸6̸2̸8̸0    (301
        4̸5̸2̸5̸2
           4
```

Diuision.

then take 1 2 out of 10, and there re‑
steth 8 : and bycause I borowed 1 of
the 3, I must cancell the 3, and write
2 ouer it, then doth the figure stand
thus.

M. Now haue you
done, and yet remay‑
neth 228, and your
quotiēt doth showe
you that yf you di‑
uide 136280 by 452, you shall fynde
your diuisor in your greater nomber
301, that is CCC. tymes and ones, &
228 remaynyng. And in the other ex‑
ample where I diuided 365 by 28, the
quotient was 13, & 1 remayned, wher‑
by I knowe that in a yere (which con
tayneth 365 dayes) there are 13 mo‑
nethes, reckenyng 28 dayes (or 4 we‑
kes) iust to a moneth, and 1 day more.
S. Why then do we call a yere but 12
monethes? M. Of that at a more con‑
uenient tyme wyll I fully enstructe
you, but now it is not conuenient to
entangle your mynde w other thyn‑
ges

Diuision.

gys thē do dyrectly pertayne to your mater, therfore yf you can remembre what you haue hard, you haue lerned a shorte maner of dyuisyon, whiche I wolde haue you often to practyse, so that you may be parfecte in it, and hereafter I wyll shewe you certayne other proper poyntes touchyng it. S. Then I pray you yet tell me how shal I exampne and trye my worke, whether I haue done well or no, that though no man be by me to tell me, yet I may perceaue it my selfe. M. Sūme men (yea and cōmenly) do trye that, by the rule of 9 as in all the other kyndes, saue that there order is this. Fyrst they cast away 9 as often as they can out of y̆ diuisor, and that that remayneth they sette at the one syde of a crosse, as in our fyrste exāple the diuisour was 28 from which you maye take 9 ones, and 1 remayneth, which they set by a crosse thus. Then do they lyke wyse exampne the quotyent (whiche in our

$\times$ 1

R   example

Diuision.

example is 13) and from thence they cast away 9 as often as they can, and the remayner they set at the other syde of the crosse, and then multyplye they to gether those 2 remayners, and to it that amounteth, they adde the remayner of the diuisiō, yf there were any, from that hole sume they withdrawe 9 as often as they can, and the rest they set at the hed of the crosse, as in our exāple, the quotiēt is 13 whiche maketh onely 4, & therfore muste you set 4 at the other syde of y̆ crosse thus. Then multiply 4 by 1 and it yeldeth but 4, therto adde the remayner of the diuisyō (whiche was 1) and it wylbe 5, whiche summe doth not amout to 9, and therfore must be set holely at the hedde of the crosse, as you see here. And this nōber on the head of the crosse is the fyrst profe

$$4 \times 1$$

$$\begin{array}{c} 5 \\ 4 \times 1 \end{array}$$

to whiche yf you fynde a nother lyke in the nomber that was diuided, then haue you done well. Therfore nowe shall

Diuision.

shall you lykewyse exampne the hole summe that was diuided, and take away 9 as often as you can, and that that remayneth, set at the fote of the crosse, and yf it be equall to that in y̆ head of the crosse, then haue you well done, els not, as in our example the hole summe was 365 whiche maketh 14, from that take 9 & there resteth 5, which set at y̆ fote of the crosse thus. And you shall see, that they agre, therfore haue you done well. S. Nowe wyll I lykewyse exampne our seconde example, where the diuisour was 452 whiche maketh 11, from thence I take 9 and y̆ 2 that remayneth, I set at the ryght syde of the crosse thus.

$$\begin{matrix} & 5 & \\ 5 & \times & 1 \\ & 5 & \end{matrix}$$

Then exampne I the quotiẽt which was 301, where I fynd but only 4 that do I set at the other syde of the crosse thus.

$$\times\ 2$$

Then do I multiply 4 by 2 and it maketh 8, to that do I adde the remayner of the diuision,

$$4 \times 2$$

K.ii. (whi=

Diuiſion.

(whiche was 228 and maketh 12) and they two make 20, wherein I fynde twyſe 9 and 2 remaynyng, that 2 muſt I ſet at yͤ hed of the croſſe thus

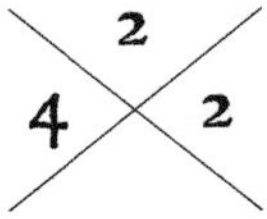

Then do I examyne the hole nomber to be diuided, whiche was 136280 where I fynde twyſe 9 and 2 remaynyng, whiche I ſet at yͤ fote of the croſſe, thus.

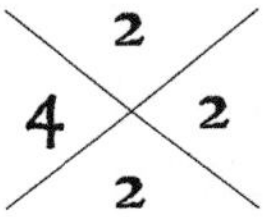

And bycauſe that it doth agre with the figure at the head of the croſſe, I knowe that the diuiſion was well wrought. M. This is the cōmen profe, howe be it, the more certayne workynge is by the contrarye kynde, as to proue Diuiſion by Multiplication, thus. Multiply the quotient by the Diuiſor, and yf the ſūme that amounteth be equall to the ſūme that ſhuld be deuided, then haue you well deuyded, els not. How be it, this muſt you marke, yͭ yf there remayned any thynge after the diuiſion, yͭ muſt you adde to the ſūme that amoūteth of the multiplication : as in our fyrſte example

Diuision.

example the quotient was 13, and the diuisor was 28. Now multiplye the one by the other, and the summe wyll be 364, to that yf you adde the 1 that remayned after the diuision, then wyl it be 365, whiche was the sūme that shuld be diuided, & therfore I knowe that I haue well done. S. Now wyll I proue the same in the second example, whose diuisor was 452, and the quotient 301 : these do I multiply together, and there amounteth 136052, to whiche yf I adde the 228 that remayned, then wyll it be 136280 which was the hole summe to be deuided, & therfore I perceaue that I haue well done. M. This is the sureste waye to exampne Diuisiō, by Multiplicatiō, and contrary wayes, the surest profe of Multiplication, is Diuision. And therfore now wyll I showe, how you may proue Multiplication by Diuision. When you haue ended Multiplication, and wolde knowe whether you haue well done or not : set ẏ grosse

K.iii.   sume

Diuision.

summe that amounteth of the multi=
plication, ouermoste, and diuide it by
the multiplyer, and yf the quotient be
the same nomber that shulde be mul=
tiplyed, then haue you well wrought,
els not, as in that example, where we
multiplyed 264, by 29 : ẏ grosse sume
was 7656. Now yf you wyll knowe,
whether that multiplicatiō be trewe,
you shall deuide that 7656, by the
multiplyer 29. And you shall percea=
ue, that the quotient wyll be 264, and
that is a token, that you haue well
wrought. S. By your pacyence I wyl
proue that, and fyrste I sette downe
the grosse sume, and the multiplyer,
not after the rule of Multiplycation,
but after the rule of Diuisiō, for now
that nomber is become the diuisor,
that was before the multiplier, I shal
set them therfore thus.   7656
And then shall I seke how   29
many tymes 2 in 7, that maye be 3 ty
mes and 1 remayneth, but then maye
not 9 be founde so often in 16, ther=
fore

Diuision.

fore must I take a lesser quotiēt, that is to saye 2, then saye I twyce 2 maketh 4, whiche I take out of 7, and there remayneth 3, then do I cancell 7 and 2, and ouer 7 I write 3, and in the quotient I sette 2, so the figures stande thus.

Then say I forth, 2 tymes 9 make 18, whiche I bate out of 36, and there resteth 18, then cancell I 3, and ouer hym set 1, and lyke wayes I cancell 6 and 9, & ouer them I set 8, so that thus stande the figures.

Then do I set forwarde the diuisor by one place, and seke a newe quotiēt, that is to say, how many tymes 2 are in 18, whiche I fynde to be 9 tymes, but then can I not fynde 9 so many tymes in 5, therfore I take a lesser quotient, as to say 8, but yet is that to greate, for yf I take 8 tymes 2 out of 18, there remayneth but 2, and I can not fynde 8 tymes 9 in 25, ther=

K.iiii. fore

Diuision.

foze yet I take a leſſer quotient, that is 7, whiche is alſo to greate, foz yf I take 7 tymes 2 out of 18, there reſteth 4, but now I can not take 7 tymes 9 out of 45, therfoze yet I ſeke a leſſer quotient, as to ſay, 6, then ſay I 6 tymes 2 make 12, that I take out of 18, and there remayneth 6, ſo I cancell that 18 and the 2, and wzite 6 ouer 8, then ſay I fozth 6 tymes 9 maketh 54, that take I out of 65, & there remayneth 11, and the figures ſtande thus.

Then muſt I ſet fozth the diuiſoz agayne, & then ſeke a newe quotient, which wyll be 4, foz thoughe I maye

```
    1
   1 6
   3 8 1
   7 6 5 6  (26
   2 9 9
       2
```

fynde 2 in 11 fyue tymes and one remayne, yet I can not fynde 9 ſo often in 16, therfoze I ſette the figures thus.

```
    1
   1 6
   3 8 1
   7 6 5 6  (264
   2 9 9 9
       2 2
```

And ỹ 4 in ỹ quotiēt I multiply into the figures

Diuision.

figures of the diuisor, sayenge, 4 ty=
mes 2 maketh 8, whiche I take out
of 11 and there resteth 3, therfore I
cancell the 11 and the 2, and set 3 ouer
ẏ fyrst place of 11, thus.
And then do I say forth
4 tymes 9 maketh 36,
which I take from 36, 7656 (264
and there remayneth
nothyng, so ẏ the quo=
tient of this diuision, where 7656 is
diuided by 29, is 264, whiche dothe
declare that yf 264 be multiplyed by
29, the sūme wyll be 7656. And thus
I perceaue nowe how bothe Multi=
plycation is proued by Diuisiō, and
Diuision also by Multiplycatiō. M.
Now haue I ended the .v. moste cō=
men kyndes of Arithmetike : for as
touchynge Mediation, Duplation,
Triplation, and such other, they are
no seuerall kyndes of Arithmetike,
but are cōtayned vnder the other, for
Mediation is conteyned vnder Di=
uision, and is nothynge els but deui

K.v.   dynge

Diuision.

dynge by 2, and so are Duplation, and Tryplation contayned vnder Multyplication, for Duplation is nothynge els but multiplyeng by 2, and tryplation is multiplyeng by 3, of whiche I wyll only propose examples, for the rules you haue heard all redye, yf you wolde medyate, or diuide into 2 this summe 4531010, you shall set 2 for the diuisor, and worke as you lerned before, as thus.

Then I fynde 2 in 4 two tymes, therfore
    4531010 (
    2

my quotyent must be 2, so I cancell 4 and 2, and remoue the diuisor forwarde thus.

Thē agayne I fynde 2 in 5 twyce, & 1 remay=
    4̸531010 (2
    2̸2

nynge, so I write 2 agayne for my second nomber of the quotyent, and cancel 5 and 2, and ouer 5 I set 1 thus.

Then remoue I the
dyuysor forwarde,
and seke a newe quo
        1
    4̸5̸31010 (22
    2̸2̸

tyent, whiche is 6 then say I 6 tymes
                      2 make

Diuision.

2 make 12 take that out of 13, and ther resteth 1, so I cancel 2 and 13, and ouer 3 I sette 1, thus.

Then remoue I the
diuisor forwarde, &
~~4~~~~5~~~~3~~1010 (226
~~2~~~~2~~~~2~~

seke a newe quotient which is 5, then take I twyce 5 out of 11, and there restethe 1, so I cancell the 2 & the last fygure of 11 & let the fyrst stand thus.

Then remoue I the
diuisor forwarde &
seke a newe quotyēt
~~1~~~~1~~
~~4~~~~5~~~~3~~1010 (2265
~~2~~~~2~~~~2~~~~2~~

whiche is 5 then take I 2 fyue tymes out of 10 and there resteth nothynge, then remoue I agayne the diuisor forwarde thus.

But bycause I
can not fynde the
~~1~~~~1~~
~~4~~~~5~~~~3~~~~1~~010 (22655
~~2~~~~2~~~~2~~~~2~~~~2~~~~2~~

diuisor in the nomber ouer it, I must set a cypher in the quotyent, & remoue the diuisor to the nexte place thus.

Then seke I a
newe quotyent
whiche I fynd
~~1~~~~1~~
~~4~~~~5~~~~3~~~~1~~010 (226550
~~2~~~~2~~~~2~~~~2~~~~2~~~~2~~

to be 5, for so many tymes may I haue

2 in

Diuision.

2 in 10. Thẽ haue I fully ended this Mediatiõ, oꝛ Diuision by 2, and the quotient is this 2265505, whiche is the halfe of 4531010, as you may trye by Duplation: foꝛ double that quotient, oꝛ multiply it by 2, and the same nõber wyll amoũt. But I wyll no lenger tary about these, seynge they are but membꝛes of the other kyndes.

Easy formes. But here nowe wyll I teache you certayne easy foꝛmes bothe of Multiplication and of Diuision, and fyꝛste of Multiplication. Yf you wolde therfoꝛe multiplye any summe by 10, you shal nede to do no moꝛe but adde a cipher befoꝛe his fyꝛst place: as foꝛ example, 36 multiply by 10, make 360. Lyke wayes yf you wyll multiplye any summe by 100, put two ciphers at his begynnynge. So yf you wolde multiply any sũme by 1000, adde thre ciphers to the begynnynge of it. S. This do I well perceaue, and also the reason of it. M. I wyll omytte all reasons tyll our nexte metynge, when I

shall

Diuision.

shall tell you the reasons of all other partes of Arithmetike also : and as to our mater now, loke (as I haue tolde you) that you do bothe remembre it, and also often practise it. But yf you wold multiply any nōber by 5, marke fyrste whether the nomber be euen or odde, and yf it be euen, take the halfe of it, and write a cypher at the begyn= nynge of it, as for example : I wolde multiply 2564 by 5, I take the halfe of it, which is 1282 (as you may know by Mediation) and before it I sette a cypher thus, 12820, and this is 5 ty= mes 2564. And thus may you do with any other euen sūme, that you wolde multiply by 5. But yf the summe be odde, as for example : 2563, then must you take the lesser half of it, or (if you wyll) take awaye 1 from the fyrste fi= gure (as here from 3) and then take the halfe of the reste, and set before it 5, as of 2563, the lesser halfe is 1281, for here I take but 1 for the halfe of 3, and yf I putte 5 before that lesser

halfe

Diuision.

halfe, then haue I multiplyed it 5 ty=
mes, as thus, 12815. S. What meane
you by the lesser halfe? M. There is
no iuste halfe of any odde nōber, ther
fore yf we diuyde an odde nomber in=
to two partes as nygh equall as can
be, yet wyll the one halfe exceade the
other halfe by one, as for example :
The two most nerest halfes of 9, are
5 and 4, and lykewayes of 15, are 7 &
8, where you se, that the one parte
styll is greater then the other by one.
Now it is easy to knowe which is the
greater halfe, & which the lesser halfe.
S. Then I perceaue you, and can do
lyke wayes I doubt not w̄ any sūme :
for yf it be not very easy to parte into
halfe, then wyll I do it by Mediatiō
easely ynoughe. M. That is a sure
way. And now haue you learned how
to multiply easely by 5, 10, 100, 1000,
and of lyke maye you do with any o=
ther of ẏ sorte. But nowe yf you wyll
multiply by 20, 30, 40, and so forth :
or by 200, 300, and suche lyke, where
there

Diuision.

there is one cypher in the fyrst place, or many orderly in the fyrste places, you shall take awaye those cyphers and multiplye the summe only by the other figure, or figures (yf they be many) and then at the begynnyng of the sume that amouteth shall you set so may cyphers as you toke awaye. Example of 2873, which I wold multiply by 300. Fyrste I caste awaye ye two ciphers from the multiplyer, and I multiply the sume by onely 3, that is lefte, and it amounteth to 8619, before whiche I put the two cyphers, yt I toke awaye before, and then is it 861900. And that is the sume that amounteth, when 2873 is multiplyed by 300. S. And yf there were two or moze figures besyde the cyphers, I must onely take awaye the cyphers, and multiply by the other figures, as I learned before, as yf I wolde multiply 93648 by 25000, I shulde take away the thre cyphers, & multiply ye sameby 25, & then at ye begynnyng of

that

Diuision.

that totall summe shulde I adde the thre cyphers agayne. M. Euen so, but and yf it chaūce the nōber that shulde be multiplyed, or bothe the summes, as well the nomber that shuld be multiplyed, as the multiplyer, to haue cyphers in theyr fyrst places, euermore cast away the cyphers, and worke by the reste: but remembre to restore as many cyphers to the amoūtyng sūme as you bated before, as in this example: 30200 shal be multiplyed by 206, I shall onely take awaye the two cyphers from the greater nombre, and then multiply 302 by 206, and afterwarde adde the two cyphers agayne. But yf I wolde multiplye the same 30200 by 2060, I shall not only take awaye the two cyphers from the nōber that shulde be multiplyed, but also I may take awaye the one cypher from the multiplyer, and then must I adde 3 cyphers to the summe that amounteth, but take hede that you take away no cypher that commeth
after

Diuision.

after any fygnyfyenge fygure, as in this laſt exſample, you may not take away that in y̌ .iiii. place of y̌ hygher nomber, nother that in the .iii. place of the multyplyer, how be it yet this you may do, yf one cypher oʒ moʒe come in the myddes of your ſummes you may multyply by the other figu=res, and ouerſkyp them, but ſo, that you geue euery figure his dewe place as thus, I wyll multiplye 3026 by 2004, therfoʒe I ſet them thus.

```
  3 0 2 6
    2 0 0 4
  ─────────
      2 4
  1 2 0 8
```

And thus I multiply them, fyʒſt 4 tymes 6 make 24 I ſet the 4 vnder y̌ fyʒſt place and kepe the 2 in my mynde, oʒ wʒyte it downe foʒ eaſye remembʒaunce, then ſay I agayne 4 tymes 2 maketh 8, and 4 tymes 0 ma=keth 0, then 4 tymes 3 maketh 12, but now when I come to the nexte cypher bycauſe that it multiplieth nothyng, I let it go, and lyke wayes the ſecond cypher, but then when I do come to the 2, and multiply it into the 6 of the

L. ouer

Diuision.

ouer nomber I must take hede (accor=
dyng as I taught you in Multipli=
cation) that the fyrst nomber amoun=
tynge of the Multiplication, be set
vnder the multiplier ryght, and the
other orderly toward the lefte hād, ac=
cordyng as you may se in this erāple.
Where yf you had expres=   3 0 2 6
sed the cyphers, after the   2 0 0 4
comen rate, then shuld the   2 4
fygures stande thus.   1 2 0 8
         3 0 2 6 But in effecte 1 2
         2 0 0 4 al is one, saue 604
            2 4 that the fyrst waye by o=
       1 2 0 0 uerscypppynge of the cy=
       0 0 0 0 phers is the shorter and
       0 0 0 0 easyer waye, for that in
          1 2 effecte they be bothe one
604   the Addition of the par=
cells wyll declare, which in both wyll
appere thus, 6064104, and nowe
wyll I make an ende of this mater. S.
Syr I thanke you, for I se greate
ease in this wayes of Multiplication
and if you can shewe me suche lyke in
                                        Diui=

Diuision.

Diuision, you shall greatly forther me. M. Yes I wyll teache you summe easye wayes in Diuision also, & fyrste this, yf you wolde deuide any summe by 10 you shall only with your penne make a square lyne betwene the fyrste fygure of your summe, and the second and then haue you done, for the hole nomber that foloweth that lyne, standeth for the quotyent, and the fygure that is before the lyne is the remayner, as for example 3648, diuided by 10 wyll stand thus.   364|8

Where 364 is the quotyent, and betokeneth that so many tymes are 10 in 3648 and the 8 after ẏ lyne is the remayner, whiche can not be diuided into 10 but by breakynge it into fractions wherwith I wyl not medle yet, and so lyke ways yf you wold deuide any summe by 100, with your penne you shal cut away the 2 fyrst fygures and yf you wold deuyde by 1000, you must cut away the 3 first fygures & so of any other diuisor whose last figure

L.ii.   is

Diuision.

is 1 and ye other be cyphers, loke how many cyphers the deuysor hath, and so maney fygures at the begynnynge shall you cut away with the squyer lyne, and they stande all wayes for the remayner, bycause they are lesse then the deuysor, and can not be deuided by it, and the other fygures that be behynd the lyne stand for the quotiente: But now yf your deuisor haue any other fygure in his laste place then 1 and in al his other places haue cyphers, loke howe manye cyphers there be, cutte awaye so many of the fyrst fygures of ye nombre that shulde be deuyded, and deuyde the reste that foloweth ye lyne, by that fygure that is in the laste place, as yf it were the hole deuysor: example of 64284 whiche I wolde deuyde by 300, here must I cutte away the 2 fyrste fygures (for so many cyphers my diuisour hath) and must diuide the reste by 3, whiche is the figure in the laste place of the diuisor. Fyrst therfore I parte
away

Diuision.

awaye the two fyrst figures, and the sume standeth thus, 642|84. Then do I diuide 642 by 3, and the quotiēt is (214, for in 6 I fynde twyse 3, and in 4 ones, and 1 remaynyng, which 1 with the 2 next before doth make 12, wherin I fynde 3 foure tymes, and this is a reddye waye to turne shyl=lynges into poundes: for syth 1 poūd doth cōtayne 20 shyllynges, I muste diuide the hole nombre of shyllynges by 20, therfore easely to do it, I se y͏̄ my diuisor hath 1 cypher, and ther=fore I cut awaye one figure from the begynnyng of the hole summe of shyl lynges, and then I do mediate or di=uide by 2 the other figures or summe that foloweth. S. I wyll put an exam=ple: Yf I wolde diuide 64287 s̄. by 20, that is to say, yf I wolde turne so many shyllinges into poūdes, I must cut awaye the fyrste figure, that is 7, and diuide the reste, that is 6428 by 2, so shall the quotient be (3214, wher by I knowe, y͏̄ 64287 s̄. make 3214 li,

L.iii. and

Diuision.

and 7 s. remaynyng. M. Now proue by multiplycation whether you haue well done or no. S. The quotient is 3214 whiche I do multiply by the diuisor 2, and it doth amoūte to 6428. M. Hereby may you perceaue not onely that you haue well done, but also how by diuision you may turne shyllynges easely into poundes. And contrary wayes, by Multiplication you maye turne poūdes into shyllynges. But here shall you se amōgest diuers men diuers fourmes of such diuision, but yf you marke what I haue tolde you, you shall perceaue easely all theyr wayes: for some men do not cut away so many of ye fyrst figures, of the sūme that they wolde diuide, as there are cyphers in the fyrste places of theyr diuisor, but they set all theyr cyphers orderly vnder the fyrste places of the nomber that they wolde diuide, and then with the other figure (or figures yf they be many) they diuide the reste of theyr sūme. Example. Yf they wold diuide

Diuiſion.

diuide 725931 by 3400, they ſet theyr ſūmes thus. 725931
And then do they diuide 34 00
orderly tyll they cõme to the cyphers, for there they ſtay & ende theyr worke: as in this example: They ſeke howe often 3 maye be founde in 7, whiche is 2 tymes, and 1 remaynyng, therfore they ſet 2 in the quotient, and cā cell 3 and 7, and ouer 7 they ſet that 1 y̌ remayned, thus.

```
       1
    7̸25931 (2
    3̸4  00
```

Thenne do they go forth ſayenge, 2 tymes 4 maketh 8 whiche they take out of 12, and there remayneth 4, thus.

Then renewe they the diuiſor forward, and ſeke how often 3

```
      1̸4
    7̸25931 (2
    3̸4̸  00
```

may be found in 4, which is but ones and 1 remayneth, then ſet they 1 in y̌ quotient, and cancell 3 and 4, & ouer them they ſette that 1 thus.

L.iiii. Then

Diuifion.

```
            1
          1 4
         7 2 5 9 3 1  (21
         3 4 4   0 0
            3
```

Thẽ take they ones
4 out of 15, & there
resteth 11 : oz elles
moze easely, take o=
nes 4 out of 5, and
there resteth 1, so they cancell the 4
and 5, and set 1 ouer them, thus.
Then set they fozthe
the diuisoz agayne, &
seke howe many ty=
mes 3 are in 11, whi=
che they fynde 3 ty=

```
            1
          1 4 1
         7 2 5 9 3 1  (21
         3 4 4   0 0
            3
```

mes, and 2 remaynynge, so they sette
3 in the quotient, and cancell 11 and
3, and ouer them setteth 2 thus.
Then do they multi
ply 4 by 3, whiche
maketh 12, that with
draw they out of 29,
and there resteth 17,
of which the 7 must
be set ouer the 9, &
the 1 ouer 2, thus.
And now are the 2 ci
phers nexte ensew=

```
            1 2
          1 4 1
         7 2 5 9 3 1  (213
         3 4 4 4 0 0
            3 3
```

```
            1
            1 2
          1 4 1 7
         7 2 5 9 3 1  (213
         3 4 4 4 0 0
            3 3
```

Diuision.

ynge so that the deuysor can no more be set forward, and therfore is the deuision ended, & the remayner is 1731 Now the quotient, which is 213, doth declare, that yf you deuyde 725931 by 3400, you shall fynd it therin 213 tymes, and there remayneth 1731, so shall you fynde it, yf you worke as I taught you, by cuttyng away the 2 fyrst figures, bycause of the 2 ciphers but this muste you marke (as you may perceaue by this laste example) that yf there be lefte any other remayner in the summe that was behynde the squyre lyne, that remayner must be sette to the later ende of the fyrste remayner, whiche was cut away with the square lyne, as yf you wolde deuyde 725931 by 3400, after the forme that I taught you, thē wolde your sumes appere thus. So that 17 whiche remayneth after the lyne must be sette to

```
        1
       1 2
      1 4 1 7
      7 2 5 9 3 1  (213
      3 4 4 4
        3 3
```

L.v.   the

Diuision.

the 31 (that was cutte awaye with the lyne) in hygher places, as you see here, where that 17 with ẏ 31 do make 1731. And here wolde I make an ende of Diuision, sauyng that there commeth to my mynde one late inuē= tion of easy Diuision, which I wyll brefely set forth to you, so that yf you fynde ease in it, you may vse it. By= cause that the hardest poynt in diui= sion is the reddye and easye fyndyng of the quotient nombre, and agayne yf that be truely knowen, all the reste is but lyghte to be done, therfore this wayes shall you quyckely and truely fynde the quotient. Fyrst wryte the ix fygures of nomber, I meane 1 2 3 4 5 6 7 8 9 not a longe as I haue set them nowe, but vp & downe, as in this fourme.

And at the lefte syde of thē drawe a longe lyne, as you se here : then consydre the diuisor by whiche you entēd to worke, and set it on the lefte syde of the longe lyne

	1
	2
	3
	4
	5
	6
	7
	8
	9

ryght

Diuision.

ryghte agaynst 1, and for a dystinctiō drawe a lyne vnder it, then multyplye your diuisor orderly by eche of those fygures begynnynge with 2, and so go downewarde tyll you haue ended all, and loke what doth amounte of the multiplicatiō of eche figure into yͥ dyuisor, that do you wryte agaynste the fygure wherby you dyd multiply. S. By example I may perceaue it better. M. Take this example, 263845 by 64, then must I sette the 9 figures as I sayde before, & the diuisor muste I set agaynst yͥ 1, with a lyne drawen vnder thus.

Then must I multiply that diuisor by eche figure orderly: fyrst by 2, and it maketh 128, whiche I muste sette agaynst 2 at the lefte hande. Then multiply I 64 by 3, and it maketh 192, which I set agaynst 3. Thē 4 tymes 64 make 256, that set I by 4. Then say I 5 tymes 64 make 320, that

64	1
	2
	3
	4
	5
	6
	7
	8
	9

Diuision.

that set I agaynst 5. Then 6 tymes 64 make 384, that set I agaynste 6. Then 7 tymes 64 make 448, which I set agaynst 7. Forther I say, 8 tymes 64 make 512, whiche I set by 8. And last of al I say, 9 tymes 64 make 576, which I set agaynst 9. And then they wyll stande thus.

And so is the table ended, by whiche you maye easely fynde the quotient, as you shall se by example nowe. Do you set downe the nōbers (as you lerned before) accordyng to the order of Diuisiō. S. That is thus.

64	1
128	2
192	3
256	4
320	5
384	6
448	7
512	8
576	9

263845
64

M. Now loke what nōber standeth ouer the diuisor, reckenyng therto all them that be behynde it, towarde the lefte hande. S. Then are there ouer the diuisor 263. M. That is iust. Now seke in the table, on the lefte syde, whether you can fynde 263. S. It is not there. M. Then take that nōber, y̓ is next to it be=

Diuision.

it, benethe it : J meane a leſſer nom=
ber thē 263, but of all the leſſer nom=
bers that y̆ table hath, take you that,
that goeth nyghest to 263. S. That is
256. M. So is it, & marke this euer
moꝛe, when you can not fynde iuſtely
in the table, that ſumme that is ouer
your diuiſoꝛ, then note that, that is
nexte benethe it, of any ſumme that is
in the table & loke at the ryghte hand
of the lyne, what figure oꝛ diget is a=
gaynſt that ſumme, and take that di=
gette foꝛ your quotient, & then woꝛke
on, as you learned befoꝛe : foꝛ nowe
haue J tolde you the hole vſe of this
table. How be it, yet that you maye be
ſure to vnderſtande it, J wyll ſe you
ende this enſample of Diuiſion by it.
Nowe therfoꝛe begynne agayne. S.
Fyꝛſte J ſet downe the ſūmes after y̆
cōmen maner thus,   263845
Then do J loke ouer the   64
diuiſoꝛ, and fynde there
263. Now to knowe how many tymes
64 may be taken out of 263, J reſoꝛte
                              to the

Diuision.

to the table afore sayd, and seke for $\tilde{y}$ nomber 263, but it is not there, therfore (as you bad me) I take a lesser nõber, the nexte to it, that I can fynde in the table, and that is 256, whiche nomber hath agaynst it on the ryght hande this digette 4, which I must take for the fyrste figure of my quotiết. Then do I (as I learned before) multiply that quotient into euery figure of $\tilde{y}$ diuisor orderly, withdrawynge the summe therof amountynge out of the ouer sũme, as here I saye fyrste 4 tymes 6 make 24, so I take that out of 26 sayenge, 4 out of 6 remayneth 2, which I write ouer $\tilde{y}$ 6, then 2 out of 2 remayneth nothyng, then cancell I 2 and 6, and also 6 in the diuisor, and the sũme stande thus.

$$\begin{array}{r}2\\ \cancel{2}\cancel{6}3845\;(4\\ \cancel{6}4\end{array}$$

Then do I lykewayes saye forth, 4 tymes 4 make 16, which I take out of 23, and there resteth 7 to be sette ouer 3, and that 3 with the 2 behynde it, and the 4 vnder

Diuision.

der it cancelled, as
you se here.
Then haue I done
with the fyrst figure of the quotient.
M. Now set forwarde your diuisor, &
seke a newe quotient, as you soughte
this. S. Then thus
stãdeth the figures,
so that ouer the di=
uisor I se 78, which
I seke in the table, and can not kynde
it, therfore I take the nexte lesser, and
that is 64 the diuisor it selfe. M. So
must you do whẽ there is none other.
S. Then agaynste it I kynde this dy=
get 1, which I must set in the quotiẽt
before 4, thus.
Then multyply I 6
by 1, and it is but 6
styll. M. You nede
not go about to multyplye when the
quotient is 1, for 1 doth nother mul=
tiply nor diuide, but in such case, only
subtracte the diuisor out of the nõber
that is ouer it. S. Then I take 4 out
of

Diuifion.

of 8, and there refteth 4, and 6 out of 7, there remayneth 1, ſo I cancell thoſe nombers, and wꝛite the remayners ouer theyꝛ places thus.

          1
  2̸7̸4
2̸6̸3̸8̸45 (41
  6̸4̸4
  6̸

Then ſet I foꝛwarde the diuiſoꝛ agayne, thus. Where I ſe ouer the diuiſoꝛ 144, which I ſeke in the table, and fynde it not, therfoꝛe I take ẏ nomber in the table that is nexte therto benethe it whiche I fynde to be

        1
     2̸7̸4
2̸6̸3̸8̸45 (41
  6̸4̸4̸4
  6̸6̸

128, agaynſt which in the ryght ſyde I fynde 2, which I take foꝛ my quotient, and that do I multiplye fyꝛſte into 6, and therof cōmeth 12, which I take out of 14, and then remayneth 2, that 2 I ſette ouer 4, and cācel the other figures, 1, 4, & 6, thus.

      1̸2
    2̸7̸4̸

Then ſay I foꝛth, 2 tymes 4 are 8, whiche I take out of 24, and

2̸6̸3̸8̸45 (412
  6̸4̸4̸4
  6̸6̸

there

Diuision.

there remayneth 16, of which I write the 6 ouer 4, and the 1 ouer 2, and cancel 2, 4, and 4, thus.

Now agayne I sette forwarde the diuisor thus.

And seynge ouer it 165, I seke that in the table, but fynde it not, therfore I take the nexte lesser

```
      1
    1/2
   27/4/6
  2/6/3/8/4 5 (4 1 2
   /6/4/4/4/4
      /6/6
```

which is 128, against whiche I fynde 2, ỹ do I set into ỹ quotient, & by it I multiply fyrst 6, & therof cōmeth 12, which

I take out of 16, & there resteth 4, thē cācel I 1, 6, & 6, & ouer 6 I set 4, thus.

Then do I multiply 4 by 2, and it maketh 8 whiche I take out of 45, and there remayneth 37, as in the next example folowynge.

```
      1
    1/2/4
   27/4/6
  2/6/3/8/4 5 (4 1 2 2
   /6/4/4/4/4
      /6/6/6
```

M And

Diuision.

```
    1 3
   1 2 4
  2 7 4 6 7
 2 6 3 8 4 5  (4122
  6 4 4 4 4
    6 6 6
```

And nowe haue I done. M. Well, now I se that you can worke by this kynd of Diuision, as farre forthe as I taughte you. S. Yea syr, I thanke you, and I fynde in it moche ease and certaynnesse. M. Yet one thynge I doubte whether you perceaue. What yf you dyd fynd in the table the nom=ber that standeth ouer the diuisor, what wold you nexte do? S. I thynke I shulde take the digette agaynste it on the lyfte hande, for the quotiente. M. So is it: and as often as you seke in the table and fynde your nomber iuste, the digette agaynste it is your true and iuste quotient: I call that a true quotient, yf it be the ryght quo=tient y̌ you shulde take, though your diuisor multiplied by the same, do not clearly subtracte the nomber ouer it, but y̌ there doth somwhat remayne, as it chaunced in all the examples,

that

Diuision.

that you dyd worke by : but yf it shuld chaunce (as it doth often) that your diuisor multiplyed by your quotient, do subtract cleane the nõber ouer it, then call I that quotient not only a trewe quotiẽt, but also a iust quotiẽt, bycause it doth iustly consume the nõber ouer the diuisor : and that chaunceth euer more when the nombre ouer the diuisor is iustly found in ẏ table. S. This I shall remember. M. But yet one easy poynte more I wyll tell you in this sort of Diuision, therfore marke it well : when you haue found in the table, other the same sũme that is ouer the diuisor, other the nexte benethe (for lacke of the other) then loke what diget standeth agaynst it, take that for your quotient, and bycause it is some payne to multiply the diuisor by the quotient, you shall not nede to do it, but onely take the nomber that you founde in the table, and subtract that from the ouer nomber : for yf you do multiply the diuisor by ẏ quotiẽt,

M.ii.   that

Diuision.

that wyll be the nomber that shall amounte, therfore is this waye more easyer. S. So is it, and also more certayner, for such as I am, y̌ myghte quyckely erre in multiplyenge, especyally beyng smally practised therin. M. Then proue in some brefe example whether you can do it, and so wyll we make an ende. S. I wolde diuide 38468 by 24, therfore fyrst I sette the table thus,

Then set I the two sūmes of diuision thus. 38468
And ouer the di= 24
uisor I fynde 38, whiche I seke in the table, and fynde it not, therfore take I the nexte benethe it, which the table hath, and that is 24,

24	1
48	2
72	3
96	4
120	5
144	6
168	7
182	8
216	9

the diuisor it selfe, agaynst whiche is set 1, whiche I take for the quotient, whiche I set in his place. And now I nede not to multiply the diuisor by it, but only to withdrawe the diuisor out of the 38 that is ouer it, & so remay=
neth

Diuision.

neth 14, as thus.
Then set I forwarde
the diuisor, and fynd
ouer it 144, as appereth : then seke I
that nomber in ẏ ta=
ble, and fynde it, and
agaynst it is 6, ther=
fore I set 6 before 1,
for my quotient, and I take that 144
for the iuste multiplication of the di=
uisor by that quotient, and therfore
without any newe multiplycation, I
do subtracte that 144 from the other
144, and there resteth nothynge, as
here you may se.
Therfore I set for=
ward ẏ diuisor, but
seyng it wyll not be
in ẏ nexte place (for then ouer 2 wold
be nothynge) I set it forwarde twyse,
as you se here.
And forbycause that
I coulde not set it in
ẏ nexte place folow=
ynge, therfore I sette a cypher in the

M.iii.   quo=

Diuision.

quotiēt, as you se. Then loke I ouer the diuisor, & fynde 68, whiche I can not fynde in the table, therfore take I the nexte beneth it, whiche I fynde in the table, and that is 48, & agaynst it standeth 2, whiche I take for the quotient. And then without any mul=
tiplyenge of the quotient into the di=
uisor, I do subtracte that 48 from 68, & there resteth 20, as here appereth.

```
  1 4   2 0
  3 8 4 6 8    (1602
  2 4 4 2 4
        2
```

And so haue I ended the hole diuision. M. Now can you suffici=
ently skyll in these kyndes of Arithmetike. And now for the vse of these two last, that is Mul=
tiplycatiō & Diuision, I wyll bꝛefely showe you the feate of Reduction.

☙ Reduction.

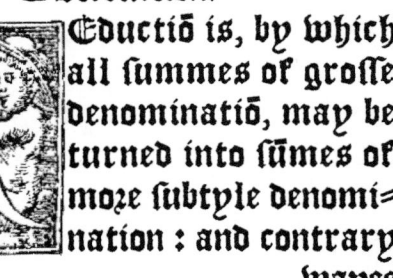Eductiō is, by which all summes of grosse denominatiō, may be turned into sūmes of moꝛe subtyle denomi=
nation: and contrary wayes

Reduction.

wayes all summes of subtyll denomi=
nation may be brought to summes of
grosser denomination. S. What call
you grosse denomination and subtyle
denominatiō? M. That I cal a grosse
denominatiō, whiche doth contayne
vnder it many other subtyler oȝ sma=
ler: as a pound in respecte to shyllyn=
ges, is a grosse denominatiō, foȝ it is
greater than shyllynges, and contay=
neth many of them. And shyllynges
in cōparisōn to poūdes, are a subtyle
denominatiō, foȝ bycause they are les=
ser then poundes, and many of them
are conteyned in one of the other: and
so lyke wayes of other thynges, what
so euer thynge is compared to other
yf it be greater, and conteyneth many
of them, it is a grosser denomination:
but yf it be lesser, so that many of thē
are in the other, then are they called
subtyle denominations: wherby you
maye perceaue, that one denominatiō
maye be called a grosse denominatiō,
and also a subtile (that is to saye, a

Grosse denominatyon.

Subtile denomination.

M.iiii. great

Reduction.

Groſſe and Subtile denomination.

greate and ſmall) in dyuers comparyſons. For ſhyllynges compared to poundes are a ſubtyle or ſmale denominatiō, but cōpared to pennes, they are a groſſe or great denominatiō. S. Now I vnderſtand the name, I pray you teache me the vſe. M. The vſe is eaſely learned, yf you remembre what you haue learned before. For yf you wyll reduce any ſume of a groſſe denomination, into a ſume of a ſmaler or ſubtiler denominatiō, you muſt conſider how many of that ſubtyler denomination do make one of the groſſer denomination, and by that nōber or numerator do you multiply the other ſume: as yf you wolde reduce 20 poūdes into ſhyllynges, you muſt conſyder, that in a pounde are encluded 20 ſhyllynges, therfore multiply the one 20 by the other 20, and there wyll amounte 400, wherby you may know that in 20 pounde are contayned 400 ſhyllynges. Lyke wayes yf you wold reduce 30 ſhyllynges into pennes, cōſide=

Reduction.

syderynge that in 1 shyllynge are 12 pennes, you must multiply 30 by 12, and it wyll be 360, wherby you kynde that in 30 shyllynges are contayned 360 pennes. And thus maye you reduce any grosse denominatiō, into a moꝛe subtyler, by multiplycation, yf you knowe how many of the lesser do make the greater: of whiche thynge I wyll anone geue you a bꝛefe table foꝛ the moste accustomed kyndes of monye, weyghtes, measures, & tyme, and such lyke, wherby you may know how often eche subtyle denominatiō is contayned in the grosser, whē you shall nede it, foꝛ the foꝛesayde kynde of reduction. And also the same shall serue you, yf you wolde reduce any summe of a subtyler denomination, into a sūme of a grosser denominatiō: Foꝛ in suche reduction you must consyder (as in the other foꝛme) how many of the smaler do make the greater, and by that nomber must you deuide the other sūme, and the quotient wyll

M.v. de=

Reduction.

declare, how many of the greater denomination are comprehended in that summe, as for example: yf you wolde knowe how many shyllynges are contayned in 3240 pennes, cosyder that 12 pennes do make 1 shyllynge, you must deuide that 3240 by 12, & your quotient wyll be 270, wherby you knowe, that so many shyllynges are in 3240 pennes. But and you wolde knowe farther, how many poundes are in those 270 shyllynges, seynge ẏ euery pound contayneth 20 shyllynges, diuide that 270 by 20, and it wyl be 13, and 10 remaynynge: wherby you maye knowe, that in 3240 pēnes, or 270 shyllynges, are 13 poūdes & 10 shyllynges: For euer more the remayner must be named by the name or denomination of the sūme that was diuided, whiche in this place were shyllynges. And thus maye you do with any other kyndes of denominations. Wherfore to the entent that you may haue a lyghte knowledge in the cō=
men

Reduction.

men coynes, weyghtes, measures, & suche other, I haue prepared here a brefe table, which shall suffyce to you at this tyme, tyll hereafter at more cōuenient oportunite, I maye enstructe you more exactly in the same.

⁋ A table for Englysh coynes. *Englysh coynes.*

⁋ A souerayne.	A quarter noble.
Halfe a souerayne.	A crowne.
A royall.	Halfe a crowne,
Halfe a royall.	A crowne.
A quarter royall.	A grote.
An olde noble.	A harpe grote.
Halfe an olde nob.	A pēny of 2 pēnes
An angell.	A dandypratte.
Halfe an angell.	A penny.
A george noble.	An halfe penny.
Halfe george nob.	A farthynge.

⁋ The valowe of Englysshe coynes.

A Souerayne is ẏ greatest Englysh coyne, and contayneth 2 Royalles or 3 Angelles, eyther 9 halfe crownes, or 4 crownes and an halfe, that is to saye 22 s̛. 6 d̛. *The valewe of Englyshe coynes.*

Halfe

Reduction.

Halfe a Soverayne is equal with a Royall.

A Royall contayneth an Angell and halfe, that is to say, 11 s. 3 d.

Halfe a Royall cōtayneth 5 s. 7 d. ob.

A quarter of a Royall, contayneth 2 shyllynges 9 d. ob. q̃.

An olde Noble, called an Henry, is worth 2 crownes, or a noble & halfe, that is 10 s.

Halfe an olde noble is worth 5 s.

An Angell contayneth a crowne and halfe or 3 halfe crownes, ẏ is 7 s. 6 d.

Halfe an Angell, is worth 3 s. 9 d.

A Noble, called a George, is worthe 6 s. 8 d.

Halfe a Noble is worth 3 s. 4 d.

A quarter of a Noble (whiche in the olde Statutes is called a farthyng) contayneth 20 d.

A Crowne contayneth 5 s, and the halfe crowne 2 s. 6 d. How be it there is another crowne of 4 s. 6 d. which is knowen by the rose syde, for ẏ rose hath no crowne over it, as in ẏ other crowne

Reduction.

crowne, but it is enuironed on the 4 quarters, w̄ 4 flowre deluces, wherby you may best knowe it. But I wyll returne to speake of the valewe of the coynes, for I entende not now to describe the fourmes of them. Nowe of golde are there no more cōmen coynes, and in syluer the greattest is a Grote, whiche conteyneth 4 pennes. Then is there an other Grote called an Harpe, which goeth for 3 d̄. Then nexte is the penny of 2 d̄. and then a dandyprate worth 3 halfe pennes. Nexte it a penny, then a halfe penny, and laste and least of all a farthynge, whose coyne is on ẏ one syde a crosse, and on the other a purculles. This I tell you, bycause I se many that can not knowe a farthynge from a small halfe penny. Nnow haue I tolde you all ẏ Englysshe coynes bothe of gold and syluer, but yet of the two moste cōmen valowers of money spake I nothynge, that is to saye, of poundes and shyllynges, whiche though they haue

Syluer coynes.

Reduction.

haue no coynes, yet is there no name moze in vse then they, of whiche the shyllynge contayneth 12 pennes oz 3 grotes : & the pounde 2 olde Nobles, 3 George nobles, oz 4 crownes, that is to saye 20 s. And this is nowe the rate of Englysshe monye. Here wolde I nowe expresse the valewes of son= dry other coynes of dyuers cōtreys, but foz thre causes I now refrayne. The fyzst & chiefest is, bycause they are not currāt by the statutes of this realme. Another cause is, by reason they are so vncertayne, that they be neuer longe at one rate. And agayne they are so dyfferent in so many pla= ces, that it were matter ynough foz a great boke to speake suffycyently of them all. How be it, yet bycause you shall not be all together ignozante of them, I wyll shewe you the valewes of some, that are most in vse : and fyzst of Fraunce. The most cōmen monye are deniers, soulx, & frances. 12 de= niers make 1 s. 20 soulx make 1 frāc, so that

Frenche coynes.

Reduction.

ſo that (as you ſe) theſe 3 kyndes are lyke in theyr rate to pennes, ſhyllynges, and poundes with vs, but that this is ẙ difference, that theyr denier is but the 9 parte of our penny, and ſo theyr ſoulx (cōmenly called ſowſes) go 9 to our ſhyllynge, and 9 of theyr frances to an Englyſhe pounde of money, ſo ẙ 3 of theyr frances make a noble. And by thoſe 3 may you practyſe how to reduce Frenche monye in to Englyſhe monye. And as for the reſte of theyr coynes I wyll omytte tyll an other tyme, when I entende to ſhowe you the rate of ſundry other kyndes of monie. But now as for the coynes of Flaunders be ſo chaungeable, that you muſte knowe them frō tyme to tyme, or elles you can not reduce them into our monye certaynly. But yet bycauſe that you ſhall haue an exāple of theyr monye to exercyſe you withall, you ſhall take thoſe that be moſte cōmen, as Stiuers, bothe ſingle and double, grotes flāmyſhe, caro=

Flaunders coynes.

Reduction.

carolus, and gyldens. A flēmyſſhe grote is lytle aboue 3 farthynges En glyſſhe. A ſyngle ſtyuer is 1 d. ob. q. The double ſtyuer is 3 d. q̄. The ſyl= uer Carolus ſyngle 2 d. q̄. q. c. The double ſyluer Carolus is 4 d. ob. q̄. q. Then is there alſo y̆ Carolus gyl= den, whiche is worth 20 ſtyuers. And the flēmyſſhe noble is worth 3 Caro= lus gyldens, and .xii. ſtyuers. But I wyll let theſe paſſe now, exhortynge you to practyſe to reduce thoſe kyn= des into Englyſſhe monye accordyng as I haue ſet forth here folowynge. 2160 deniers make 240 d̆. or 20 s̆. 3240 deniers make 360 d̆. or 30. s̆. 8352 deniers make 928 d̆. or 3 li. 17 s̆. 4 d̆. 2160 ſoulx make 240 ſhillyngs. But yf you wyll reduce Flemmyſſhe monye iuſtly, you muſt reduce it fyrſt into the ſmalleſt parte of Englyſſhe monye that is in that coyne, as for example : If you wolde reduce 368 double ſtyuers into Englyſh monye, conſyderyng that a double ſtyuer cō=
tay=

Reduction.

tayneth 3 d. q̃. you shall fyrste loke howe many q̃. be in y̆ double styuer, and you shall fynde them 13, therfore multyply the sume of the styuers by 13, and then haue you theyr valewe in farthynges, which is 4784. Now yf you deuide that by 4, then wyll there appere the nomber of p̄enes, but better it were to diuide it by 48 (for so many farthynges are in 1 shyllynge) and then wyll the quotient declare the sume of shyllynges. Lykewayes yf you wolde reduce any sume of syngle styuers into Englysh monye, you must multiply the sume fyrst by 13, & then haue you the sume of q. whiche sume yf you diuide by 8, then wyll amount the sume of pennes : or yf you diuyde it by 96, the summe of shyllynges wyll appere. But this marke in all diuision, when you do reduce to brynge one denomination into an other : yf there by any remayner after the diuision, that muste be named by the denomination of the grosse sume

N.   that

Reduction.

that was diuided: as for example: I wolde brynge 254 q̄. into penes, therfore I do diuide that 254 by 4, for so many farthynges make 1 penny, & the quotient is 63, which is the sūme of the penes, & then remayneth yet 2, which are farthyngs styll, as you may proue by diuidyng. And this must be marked in all diuisiō, namely when it is done for Reductiō. And thus moch haue I said of mony, now wyl I shew *Weyghtes.* you in like sorte y̆ distinctiō of weyghtes, after y̆ statutes of Englād, where the leaste portion of weyghte is com= *Grayne.* monly a grayne, meanynge a grayne or corne of whete, drye, and gathered out of the myddell of y̆ eare. Of these graynes in tymes passed 32 wayed iuste 1 penny of troye, and then was but 20 pennes in an vnce, but nowe are there 46 pennes in an vnce, so y̆ there are not fully 14 graynes in 1 *an Vnce.* penny. But now of vnces (after troye *a Penny of* rate, whiche is the standarde of En= *Troye.* glande) 12 do make 1 pound. But cō menly

Reduction.

menly there is vsed an other weyght called haberdyepoyse, in which 16 vnces make a pounde. Therfore when you wolde reduce vnces vnto poundes, you must consider, whether your weyghtes be troye weyghtes or haperdyepoyse, and if it be troy weight, you must dyuyde your vnces by 12, to brynge them to poundes: but yf it be haberdypoyse, you must diuide thē by 16. Now agayne there be greater weyghtes, whiche are called a hundred, halfe hundred, and quarterne, & also halfe a quarterne. &c. S. Why? so there may be rekened 20 pounde, 40 pound, 200 pounde, and such vnnumerable. M. All these are nombers of weyghte, but they haue not commen weyghtes made to theyr rate, as the other haue. And agayn these that I dyd name are not iust in nomber as they seme by theyr name: for a hundred is not iust 100, but is 112 poūd. And so the halfe hundred is 56, the quarter 28, and the halfe quarter 14.

Haberdeypoyse weyghtes.

a Hundred weyghte.

N.ii.　　And

Reduction.

Wolle weyghtes.

a Todde.

Stone.

Sacke

Chese weyghtes.

a Cloue

Wey

And this is the cōmē weyghtes vsed in most thinges ẏ are sold by weyght. How be it there are in some thynges other names: as in wolle 28 pounde is not called a quartern, but a todde: and the 14 pound is not named halfe quarterne, but a stone: & the 7 poūde halfe a stone. Other names bycause they dyffer in many places, and agree in fewe, I let them passe. But a sacke of wolle by the statutes is lymytted to be 26 stone. Now in chese though it be sold by the hūdred, & by the stone in some places, yet ẏ very weyghtes of it are cloues and weyes, so that a cloue sholde contayne 7 pound: and a wey 32 cloues, that is 224 pounde. How be it some statute bokes saye ẏ a cloue sholde be 6 pound, and some say also that a wey doth conteyne 36 cloues, and that is cōmenly vsed, for the cōmen wey is of 256 pound, that is 36 cloues, reckenynge 7 pound to ẏ cloue, & there is 4 poūd ouer weyght. And let this suffice you at this tyme, tou=

Reduction.

touchyng weyghtes. Now of weygh= tes are made other measures, bothe *Measures for* for grayne and lyquor. For a pounde *lyquor.* in weyghte, maketh a pynte in mea= *a Pynte.* sure, so that 8 pounde (or 8 pyntes) do make a galon: halfe a galon is na= *Galon.* med a pottell, & halfe a pottell is cal= *Pottell.* led a quarte, whiche contayneth two *Quarte.* pyntes. Now aboue a galon the next measure is a fyrken: then a tertian, a *Fyrken* kylderkyn or halfe barrell, and a bar= *Tertian.* rell. And by those measures are solde *Kilderkyn.* cōmely Ale, bere, wyne, and oyle, but= *Barrell.* ter & sope, salmon, herynges & eeles: but as these be vnlyke thinges, so the measure of theyr vessels do dyffer: for the measures of ale are as foloweth. *Ale measures.*

of ale ⎰ the fyrken ⎱ cōtayneth ⎰ 8 ⎱ galōs.
       ⎱ y̆ kilderkē ⎰   ⎱ 16 ⎰
       ⎱ the barrel ⎰   ⎱ 32 ⎰

of bere ⎰ the fyrken ⎱ cōtaineth ⎰ 9 ⎱ *Beere mea=*
        ⎱ y̆ kilderkē ⎰   ⎱ 18 ⎰ *surs.*
        ⎱ the barrel ⎰   ⎱ 36 ⎰ galōs.

Sope measures, bothe fyrken, kyl= *Sope mea=* derken, & barrell, shulde be all equall *sures.*

N.iii.   to

Reduction.

to Ale measures. Moreouer the sta=
tutes doth lymyte ẏ weyghte of euery
of those thre vesselles beynge empty.

A barrell ⎫   ⎧ 26 ⎫
Half barrel ⎬ to wey ⎨ 13 ⎬ pounde
A fyrken ⎭   empty ⎩ 6 ð. ⎭

Herynge. Herynges also be solde by the same
measures, that ale & sope be solde by:
but herynges also are solde by tale,
120 to the hūdred, 10 M. to the last.

Salmon and Eeles. Salmon, and eeles, haue a greater measure.

Salmō ⎧ the butte ⎫   ⎧ 84 ⎫
& Eeles ⎨ ẏ barrell ⎬ holdeth ⎨ 42 ⎬ galōs.
        ⎩ half bar. ⎪   ⎪ 21 ⎪
        ⎩ the fyrkē ⎭   ⎩ 10 ð. ⎭

Wyne measu= res. How be it, some statutes dyd lymytte
eele vessels equal tō herynge vessels.
Now as for wyne vessels seldome are
smaler then hogges heddes, whiche
are of 63 galons: euery hogges hed
is two barrelles: yet there are many
other wyne vessels, but of them all, se
this table, and marke the measures
one to an other.

Of

Reduction.

Of wine & Oyle { the rondelet, the barrell, y̆ hogges hed, the tertian, the pype, the tonne } hol= deth { 18 ð., 31 ð., 63, 84, 126, 252 } ga= lōs.

But you shall marke that there be o=ther kyndes of tertians, for there be tertians (that is to saye thyrdeles) of pypes, of hogges heddes, and of ba=rels, as well of other thynges as of wyne, also of malueseys and secke. &c. The halfe tonne is not called a pype, but rather a butte. And thus moche haue I thoughte mete to tell you at this tyme. S. And is this alwayes true? M. I haue tolde you howe it sholde be, but how it is I may not say how they do dyffer dayely from theyr iust measure: the gagers can tell you better then I. But I wyll lette this passe now, and speake brefely of the other measures, And as of weyghtes there dyd sprynge the lyquyde mea=

Tertians.

a Butte.

P.iii.   sures

Reduction.

Dry measures. ſures wherof I ſpake laſte: ſo of the ſame ſpryngeth drye meaſures, as peckes, buſhels, quarters, and ſuch like, wherby are meaſured corne and lyke graynes: alſo ſalte, lyme, coles, and other lyke. And this is the order and quantitie of them.

a Pecke. A pecke is the meaſure of 2 galons.
Buſſell. A buſhell contayneth 4 peckes.
Quarter. A quarter holdeth 8 buſhelles.
Weye. A weye contayneth 6 quarters.

Theſe are the commen names & meaſures, but in dyuers places there be dyuers ſortes. And ỹ buſhel in many places is 2 buſhels, but then is the *Stryke.* buſhel there called a ſtryke. And in ſome places halfe a quarter is called *Cornoke.* a cornoke. But theſe diuerſities are to many, to tell you brefely them all. And agayne ſyth they are agaynſte ỹ lawe and ſtatutes, I counte them vnmete to be vſed. But now remayneth *Meaſures to mete lengthe & breadthe.* yet another kynde of meaſure, wherby men mete lengthe and bredthe, & thoſe are, an ynche, a fote, and ſuche other

Reduction.

other: whole names and lengthe, this table showeth.

3 graynes of barly make an ynche. — *An Inche*
12 ynches make a foote. — *a Foote.*
3 foote make a yarde. — *Yarde.*
3 foote and 9 ynches make an elle. — *Elle.*
5 yardes and halfe make a perche. — *Perche.*
1 perche in bredthe, and 40 in length do make a rodde of londe, whiche some call a roode, some a yarde londe, and some a farthendele. — *Rodde.* *Farthendele.*
2 farthendeles make halfe an acre of ground.
4 farthendeles make an acre.

Here mought I tell you many thynges elles, touchynge measure. And also how to reduce straūge measures to our measures, but bycause it can not wel be done wout the knowledge of fractions, whiche as yet you haue not learned, I wyll let them passe tyll an other tyme, when I shall enstruct you in the pryncipels of Geometrye, wherein I shulde be enforced els to re pete the same often agayne. S. But yet

Reduction.

The partes of Tyme.
a Daye.
a Houre.

Weake.
Moneth.
Yere.

yet fyr, of the partes of tyme, I pray you tell me fomwhat. M. You knowe that a naturall day hath 24 houres, and euery houre hath 60 minutes, It nedeth not to tel you, that 7 dayes make a weke, & 4 wekes make a commen moneth, and 13 monethes make a yere, lackynge 1 daye and certayne howres and minutes: But of that I shall enstructe you here after. And here wyl I make an ende of Reductiō for this tyme, which though it be coūted no kynde seueral of Arithmetike, yet you se it is no lesse nedefull to be knowen, nor easyer to be done, then any of ẏ other. S. Mary syr it semeth vnto me moch harder then any other sorte, for it requyreth the knowledge of so many thinges. But now syr whē you se tyme I am redy to lerne forth, for as moch of Reductiō as you haue taught me I remember: but and yf I do at any tyme forgette, I shall haue recourse to ẏ tables, whiche you haue set forth for me. M. So do you, for it wyll

Progreſſion.

wyll not be remēbred wout exercyſe. Now with Progreſſiō I wyll begyn.
☙ Progreſſion

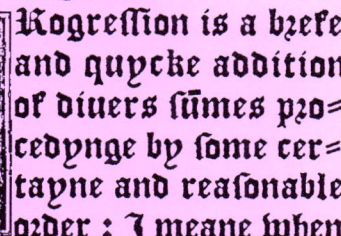
Rogreſſion is a bꝛeke and quycke addition of diuers ſūmes pꝛocedynge by ſome certayne and reaſonable oꝛder : I meane when the diſtaunce of euery 2 nombers is equall to the diſtaunce of the ſeconde nomber from the fyꝛſte. S. I vnderſtande you not well. M. By an example it wyll be playner, as here : 1, 2, 3, 4, 5, 6, here you ſe the ſeconde to dyffer from the fyꝛſte but by 1, & ſo doth all the other, one excede an other by 1 ſtyll to the ende. S. This I perceaue. M. And lyke wayes here 4, 7, 10, 13, 16, 19, 21, 24. S. Yea, they pꝛocede by ẏ dyfference of 3. M. And ſo I meane, that Progreſſion is an arte howe to adde all ſuch nombꝛes and other lyke together, moch quycklyer then by cōmen Addition, and that by this mea=
nes

Reduction.

7

nes : Tell how many nombers there are, and yf they be odde, write theyr ſume downe by it ſelfe : as in this example, 2, 4, 6, 8, 10, 12, 14, where the nombers are 7, as you maye ſe, therfore ſet downe 7 in a place alone, as I haue done in the margent here : thē adde together the fyrſt nomber and ẙ laſte, as in this exāple : adde 2 to 14, and that maketh 16, take halfe of it, and multiply by that 7, whiche you noted for the nomber of the places, & the ſume that amounteth, is the ſume of all thoſe figures added together : as in this example, 8 multiplyed by 7, make 56, and that is the ſume of all the figures. S. That wyll I proue by an other example : I wolde knowe how moche this ſume is, 5, 8, 11, 14, 17, 20, 23, 26, 29, I tell the places, & they are 9, that I note : Then I put the fyrſte nomber 5, and the laſte 29 together, and they make 34, I take the halfe of it that is 17, and multiply by 9, and it maketh 153. That you
ſaye

Reduction.

ſaye is the ſume of all the nombers. M. So ſhall you kynde it, yf you trye it. S. How ſhall I trye it? M. By Ad= ditiō: For yf you adde all the parcels together, you ſhall ſe the ſame ſume amount, yf you dyd worke well: and that Additiō trieth all kyndes of Pro greſſion. S. Then I can adde by Pro greſſion, yf the nombers of the par= cels be odde. But what yf they be euen? as in this example, 1, 2, 3, 4, 5, 6, 7, 8. M. When the nomber of the parcels be euen, then note that alſo, as you dyd before, & lyke wayes adde the fyrſte ſume to the laſte, and by the halfe of the nomber of the places do you multiplye it, as in your example, the parcels are 8, that note I, then addynge the fyrſte ſume to the laſte, there amounteth 9, that do I multy= ply by the halfe of parcels, that is by 4, and it maketh 36, which is ẏ ſume of the 8 parcels. But yf you wyll take one rule for theſe both, do thus. Mul tiply the halfe of the one by the other
hole

Progreſſion.

hole, and the ſūme wyll amounte all one. For ſomtyme it chaunceth that the parcelles be odde, ſo that theyr halfe can not be taken: and ſomtyme it chaunceth the addition of the fyrſte nomber and the laſte to be odde, ſo ẏ the halfe of it can not be taken, but they wyll neuer be both odde. S. Thē I perceaue this, yf there be no more longynge to it. M. This is ynough for Progreſſion Arithmeticall, how be it, there is another maner of Progreſſion called Geometricall, when the nombres encreaſe by a lyke proportion, that is, yf the ſeconde nomber contayne the fyrſt 2, 3, or 4 tymes and ſo forth, then the thyrde contayneth the ſeconde ſo many tymes alſo, and ſo the fourthe the thyrde, and the fyfte the fourth: wherfore I ſet thoſe thre examples. Here in the fyrſt example you ſe, that euery nomber contayneth the other, (that goeth nexte before hym) 2 tymes

Progeſſion Arithmetycall.

Geometricall.

3, 6, 12, 24, 48,
1, 3, 9, 27, 81,
2, 10, 50, 250,

Progreſſion.

mes: and in the ſecond example 3 ty=
mes: in the thyrde example 5 tymes.
Nowe yf you wyll knowe howe to
fynde eaſely the ſumme of any ſuche
nombers, do thus: conſyder by what
nomber they be multiplyed, whether
by 2, 3, 4, 5, or any other, and by the
ſame nōber do you multiply the laſte
ſūme in the Progreſſion. S. I praye
you worke it by this example, 2, 8, 32,
128, 512, 2048, whiche ſumme I haue
encreaſeth by the multiplycation in=
to 4. M. Then muſt I multyply the
laſte ſūme (whiche is 2048) by 4 alſo,
and it wylbe 8192. Now muſt I bate
from this ſūme the fyrſt nombre of y̶
Progreſſion, whiche here is 2, then
reſteth 8190, which ſume I muſt diuide
by 1 leſſe then was the nomber that I
multyplyed by. Seynge then I mul=
tiplyed by 4, I muſte diuide by 3, ſo
diuidynge 8190 by 3, y̶ quotient wyll
be 2730, which is the ſūme of all the
Progreſſyon. And now to proue whe=
ther you can do the ſame, I geue you
theſe

Reduction.

these nombers to adde, 3, 15, 75, 375, 1875, 9375, 46875. S. I can not wel tel by what nōber this progression doth encreace. M. In any such doubte do thus: Diuide the seconde nomber by the fyrste, and the quotient wyll show you the nomber that engendreth the progression. S. Then is that nomber in this example 5, for so many tymes is 3 in 15. M. So is it. Now worke as I taught you. S. The last nombre is 46875 whiche I multyply by 5, and it yeldeth 234375, frō whiche I bate the fyrst nombre of y̆ progression that is 3, and there resteth 234372, which I diuide by 4 (for that is 1 lesse then 5) & the quotient is 58593, which is the hole sūme of the Progression. M. Now yf you remēbre wel this, you haue learned the arte of Progressyon bothe Arithmeticall, and also Geometricall: whiche you maye proue other by Subtractiō of eche nombre alone from the sūme, and so wyll there no=
thynge remayne, other by addyng to=
gether

Progression.

gether of all the parcelles, for so wyll the same sūme amount. And now for the vse of this rule I wyll put forthe to you certayne questyons, whiche some do referre to Addition, but not so iustly, as I do vnto this rule of Progression: and some as vncircum=spectly referre the same to Duplatiō.

The fyrst question is this.

¶ Yf I solde vnto you a horse, ha=uynge 4 shoes, and in euery shoe 6 nayles, with this condition, that you shall pay for the fyrste nayle 1 ob. for the second 2 ob. for the thyrde 4, and so forth doblyng vnto the ende of all the nayles. Nowe I aske you howe moche wolde the hole pryce of the horse come vnto. S. Fyrst to knowe the nom ber of the nayles, I must multiply 6 by 4, and that maketh 24, then I wyll do thus. I wyll write the nomber of the nayles euery one in order from 1 to 24, and agaynste eche nombre of nayles the sūme of halfe pēnes dewe, as the order of Duplacion teacheth,
      D.   and

Questiō of sel=lyng of a horse.

Progreſsion.

and in this figure appeareth. Then do I reſorte to the rule of Progreſsion, where I conſyder, that the encreaſe of this ſũme procedeth by multiplycation of 2, and therfore I do multiply p̃ laſt ſũme by 2 alſo, & it yeldeth 16777216, from whiche I abate the fyrſte nõber, whiche is 1, and then reſteth 16777215, which I ſhulde diuide by 1 leſſe, then I dyd multiply. But ſeynge that it is 1, I nede not to diuide it, for 1 (as I haue before ſayde) doth nother multiplye

1	1
2	2
4	3
8	4
16	5
32	6
64	7
128	8
256	9
512	10
1024	11
2048	12
4096	13
8192	14
16384	15
32768	16
65536	17
131072	18
262144	19
524288	20
1048576	21
2097152	22
4194304	23
8388608	24

nor

Progreſſion.

nor diuide: therfore I take that ſumme 16777215, for the hole ſumme of halfe pennes, whiche by reduction I fynd to be 699050 s̷. and 7 d̷. ob. that is 34952 li. 10 s̷. 7 d̷. ob. M. That is well done: but I thynke you wyl bye no horſe of the pryce. S. No ſyr, yf I be wyſe. M. Well, then aunſwer me to this queſtyon.

A lorde delyuered to a brycke layer, a certayne nomber of lodes of bricke, wherof he wylled hym to make 12 walles, of ſuch ſorte that the fyrſte walle ſhulde receaue 2 thyrdeles of the hole nomber: and the ſeconde 2 thyrdeles of that, that was lefte: and ſo euery other 2 thyrdeles of that that remayned: and ſo dyd the brycke layer. And when the 12 walles were made, there remayneth 1 lode of brycke. Nowe I aſke you how many lode went to euery wall, and howe many lode was in the hole? S. Why ſyr? it is vnpoſſyble for me to tell. M. Nay, it is very eaſy yf you marke it wel: Marke well, y̷ I

Queſtyon of brycklaynge.

O.ii.   ſayd

Progreſſion.

ſayde that euery wall ſhulde receaue 2 thyrdeles of ẏ ſũme that was lefte. Now take awaye 2 thyrdeles from any ſumme, & you muſt nedes graũt, that that which remayneth, is 1 thyr= dele of the ſũme laſte before: example of 9, from which yf you take 2 thyr= dels, there wyll remayne 3, which is 1 thyrdele of 9: lyke ways from 3 bate 2 thyrdeles, and there wyl remayne 1. S. This is true, and now I perceaue that the leaſt wall had but 2 lode of brycke. M. And by the ſame reaſon may you know how many lode euery wall had, accordynge as this figure doth ſhowe, & lyke wayes what ẏ hole ſũme of bryckes was: for yf you make 12 ſummes multiplyenge by 3 ſtyll, from the laſt remayner, as you ſe here on the lefte ſyde of ẏ table, there wyll appere all the remayners after euery wall: and yf you multyply the laſte of thoſe 12 ſummes by 3 alſo, then wyll that be the ſumme of the lodes, which were delyuered to the brycke layer.

Agayne

Progreſſion.

1	12	2
5	11	6
9	10	18
27	9	54
81	8	162
243	7	486
729	6	1458
2187	5	4374
6561	4	13122
19683	3	39366
59049	2	118098
177147	1	354294
	531441	

Agayne, yf you do double euery re=
mayner, as you ſe at the ryght ſyde of
this table, thoſe nombers wyll ſhow
the ſumme of lodes that went to eche
wall : wherby alſo you may perceaue,
that eche wall was 3 tymes ſo greate
as the nexte leſſer. S. Lo, nowe it ap=
peareth eaſy ynoughe. Now ſurely I
ſe that Arithmetike is a ryghte excel=
lent arte. M. You wyll ſaye ſo when

D.iii.   you

Progreſſion.

you knowe more of the vſe of it: for this is nothynge in compariſon to other poyntes that maye be wroughte by it. S. Then I beſeche you ſyr, ceaſe not to inſtruct me farther in this wōderfull connynge. M. By the order of the ſcience (as men haue taughte it) there ſhuld folowe nexte, the Extraction of rootes of nomber, whiche bycauſe it is ſomwhat harde for you, yet I wyll let it paſſe for a whyle, & wyll teache you the feate of y͏̈ rule of Proportions, whiche for his excellency is called the Golden rule. Whoſe vſe is by 3 nombers knowen, to fynde out another vnknowen, which you deſyre to knowe: as thus. If you paye for your borde for 3 monethe 16 s. how moch ſhall you paye for 8 monethes. To knowe this & all ſuche lyke queſtions, you ſhall conſyder which two of your 3 nombers be of one denomination, and ſet thoſe 2, the one ouer the other, ſo that the vndermoſt be it, that the queſtiō is aſked of: As in my queſtiō

The rule of proportyōs called the Golden rule.

Queſtiō of bordynge.

Progreſſion.

queſtion 3 and 8 be bothe of one de
nomination, for they bothe be mone=
thes : and bycauſe 8 is the nomber y�
the quotiēt is aſked of, I ſet then one
ouer y̆ other, and 8 vndermoſt, thus,
with ſuch a croked draught of lynes.
3 —— Then do I ſet the other nōber,
8 —— whiche is 16, agaynſt 3 at the
ryghte ſyde of the lyne, thus. 3 —— 16
And now to knowe my que= 8 ——
ſtyon, thus muſte I do : I muſt mul=
typly the lowermoſt on the lefte ſyde
by that on the ryghte ſyde, & the ſūme
that amounteth I muſt diuide by the
hyeſt on the lyfte ſyde : Or in playner
wordes thus : I ſhall multiply the nō
ber, of whiche the queſtyon is aſked
(whiche is called the thyrde nomber) *The Thyrde nōber.*
by the nomber of an other denomina=
tion (whiche is called the ſecond) and *the Seconde.*
that ſūme that amounteth, muſte I
diuide by the ſumme of lyke denomi‐
nation, whiche is called the fyrſte. *the Fyrſte.*
Then for the knowledge of this que=

D.iiii.   ſtion

Progreſsion.

ſtion, I multyply 8 into 16, & there amounteth 128 which I diuide by 3, and it yeldeth 42 s̷. and 2 s̷. remayneth, which I turne into pennes, & they be 24 d̷. of which the thyrd part is 8 d̷. ſo the thyrde parte of 128 s̷. is 42 s̷. 8 d̷. whiche ſume I wryte at the ryght hande of yͨ figure agaynſte 8, thus.

                 3   16 s̷.
                       Z
And hereby I know, 8   42 s̷. 8 d̷.
that, yf 3 monethes borowyng do come to 16 s̷. that 8 monethes borowynge wyll come to 42 s̷. 8 d̷. & lyke wayes of any other lyke queſtyon. But here muſte you marke, that the fyrſt nomber and the thyrde be of one denomination, and alſo the ſeconde and the fourthe, for whiche you ſeke : or elles be of ſuche denominations, that you in workynge maye brynge them into

Queſtiō of ex=
pences.

one : As yf a man ſhulde aſke me this queſtion : 12 weekes iorneyng coſte me 14 nobles, how many poūdes is that in one yere? Here you ſe no 2 nōbers of one denomination, but yet in

wor=

Progreſſion.

workynge you maye turne them into lyke denominations, as thus: Turne the one yere into 52 wekes, and the fourth ſūme wylbe nobles, by the or= der of the workynge. Then to knowe this queſtion, multiply ẏ thyrde ſūme 52 by the ſeconde 14, and the ſumme wylbe 728, that diuide by 12, and it wyll be 60, and 8 remaynyng, which yf you turne into ſhillynges they wyl be 53 s̃. 4 d̃. whiche yf it be diuided by 12, wyll yelde 4 s̃. 5 d̃. ¢ the thyrd parte of a penny: put this 60 nobles (which maketh 20 li) with the 4 s̃. 5 d̃. ¢ q̈. and lytle more: for the ſūme that anſwereth to the queſtyon and it is ẏ expenſe of a yere, and the ſūmes wyll be thus.   12   14 nobles.
And take this   52   60, 4 s̃. 5 d̃. q̈. for a generall rule, that euermore the thyrde nomber be it, that the quotient is ioyned with: and the fyrſte, the nō ber that is of the ſame denominatiō, then muſt the ſeconde nedes be that other. And remēbre alſo that ẏ place

A general rule.

D.v.   of

Progreſſion.

of the fyrſte nomber is the hygheſt on the lefte ſyde: and the place of the ſeconde ryght agaynſt it on the ryghte ſyde: the place of the thyrde nomber is vnder the fyrſte, as by thoſe examples you haue ſene. S. This I truſte I can do. M. But and the queſtyō be aſked thus: In 8 wekes I ſpende 40 s. howe longe wyll 105 s. ſerue me? Though the order ſeme vnlyke, yet take you 105 for the thyrde nomber, and 40 beynge of the ſame denomination for the fyrſt, and thē 8 for the ſeconde. Then multiply 105 by 8, and it wyll be 840, whiche yf you diuide by 40, it wyll yelde 21, whiche is the fourth nomber, & ſheweth how many wekes 105 s. wyll ſerue, yf you ſpende 40 s. in 8 wekes. The figure of this queſtyon is this, as yf you ſhuld ſay: yf 40 s. ſerue for 8 wekes, 105 ſerue for 21 wekes. Other diuerſites there be of workynge by this rule, but I had leauer that you wolde learne this one well,
then

Queſtiō of expences.

$$\begin{array}{c} 40 \diagdown 8 \\ 105 \diagup 21 \end{array}$$

Progreſſion.

then at $\tilde{y}$ begynnyng to trouble your mynde w̃ many formes of workyng, ſyth this way can do as moch as al $\tilde{y}$ other, and here after you ſhall learne the other more conueniently. But yet before we make an ende of this rule, this ſhall you note, that there is ano= ther order quyte contrarye to this $\tilde{y}$ you haue learned. For in this rule hetherto, euermore loke howe moche the thyrde nomber is greater then the fyrſte, ſo moche the fourth nomber is greater then the thyrde: and contrary wayes, loke how moch the fyrſt ſume is greater thē $\tilde{y}$ thyrd (yf it do chaūce ſo) ſo moche is the ſecond ſume grea= ter then the fyrſte. But there is a cō= trary order, as this: That the grea= ter the thyrde ſume is aboue the fyrſt, the leſſer the fourth ſumme is beneth the ſeconde: and this rule you maye call the backer rule: as in example. If I haue bought 30 yardes of cloth, of 2 yardes bredthe, and wolde bye canwas of 3 yardes brode to lyne it

with

The Backer rule.
Queſtiō of by enge of clothe.

Progreſſion.

with all, how many yardes ſhulde I nede? S. Why? there is none ſo brode. M. I do not care for that, I do putte this example onely for your eaſy vnderſtandyng: For yf I ſhulde put the example in other meaſures, it wold be harder to vnderſtand: but now to the mater. Yf you wyll knowe this queſtyon, ſet your nōbres as you dyd before: but you ſhall multiplye now the fyrſte nombre by the ſeconde, and that aryſeth therof, you ſhall diuide by the thyrde, which thyng yf you do here, I meane, yf you multyplye 30 by 2, it wyll be 60: which ſume yf you diuide by 3, there wyll appere 20, wherby I knowe, that yf 30 yardes of clothe of 2 yardes brode ſhulde be lyned w̄ canwas of 3 yardes brode, 20 yardes of canwas wolde ſuffice, as this figure ſhoweth alſo.

2 — 30
3 — 20

And now bycauſe you founde faulte at my exāple, how ſay you perceaue you this? S. Yea ſyr, I ſuppoſe. M. Thē anſwer me to this que=

Progreſſion.

queſtion : How may elles of canwas of elle bredthe, wyll ſerue to lyne 20 yardes of ſeye, of 3 quarters of a yard brode? S. In good fayth ſyr I can not tell, for I knowe not howe to brynge the ſumes to lyke denominatiōs. M. Then I wyll tell you, ſyth there is mention here of quarters, & agayne euery one of the meaſures both elles and yardes may be parted into quar=ters, do you parte them ſo, bothe in y̌ bredthe & lengthe, and then put forth the queſtyon by quarters. S. Then I ſhal ſay thus : how many quarters of canwas of 5 quarters brode wyl lyne 80 quarters of 3 quarters brode? M. Now anſwere to the queſtyō. S. Fyrſt I wyll ſet them downe in theyr forme thus : for 5 is ioyned with   3 — 80 the queſtyon, and is ther=   5 fore the thyrde nomber, then is 3 the nomber of the ſame denomynation, I meane bycauſe they be bothe referred to bredthe. Now I multiply 80 by 3, and it is 240, whiche I diuide by 5,

and

Progreſſion.

and it yeldeth 48. Then saye I, that 48 quarters of 5 quarters brode, wyl ſuffyce to lyne 80 quarters of 3 quarters brode. M. Turne the quarters agayne into elles and yardes. S. Thē I say, ethat 9 elles and 3 quarters of a yarde of elle brode, wyl ſerue to lyne 20 yardes of 3 quarters brode, as this figure sheweth.

$$3 \diagup 80$$
$$5 \diagdown 48$$

M. This rule is ſo profitable for all eſtates of men, that for this rule onely (yf there were no more but it) all men were bounde hyghly to eſteme Arithmetike. By this rule maye a capitayne in warre worke many thynges, as I wyll hereafter enſtructe you abundauntly: onely now I wyll ſhowe you this one exāple. *Queſtiō of prouiſyō, to chynge an armye.* Yf it ſhuld chaunce a capitayne, whiche hath 40000 ſowldyours to be ſo encloſed w͡t his ennemyes, that he could haue no freſhe purueyaūce of vptayles, and that the vptayles whiche he hath, wold ſerue that armye but only 3 monethes, how many men ſhuld he dymyſſe, to make that vptayle to ſuf=
fyce

Progreſſion.   **106**

fyce the reſydue 8 monethes? S. As you taughte me, I ſet ẏ nōbers thus: 3 — 40000 ſayeng: Yf 3 monethes 8 ſuffyce 40000, to how many wyll 8 monethes ſuffyce? To know this, I multiply ẏ fyrſt nōber 3 ito ẏ ſecōd 40000, & it yeldeth 120000, which ſumme, I diuide by 8 & there wyll be in the quotiēt 15000 whiche yf I do ſubtracte frō 40000, ẏ remayner wyll declare ẏ he muſt dimiſſe 25000, as this figure ſheweth. 3 — 40000 M. Well, ſyth you per= 8 — 15000 ceaue now the vſe of this rule, I wyll ſhowe you other, which enſewe of the ſame. And fyrſte the double rule, whi= *The Double* che is ſo called, bycauſe there is in it *Rule.* double workynge, by whiche thynge onely it dyffereth from this. S. Then by an example I ſhall vnderſtande it well ynough. M. So ſhall you: and let this be the exāple. Yf the caryage of 100 pound weyghte 30 myles, do coſte 12 d̃. how much wyll ẏ caryage of 500 weyghte coſte, beynge caryed 100 myles? S. I praye you ſhowe me the

Progreſſion.

the workynge of it. M. You muſte make two workynges of it: the fyrſte thus. Yf 100 pound weyghte coſt 12 d, how moch wyll 500 pounde coſte? Set your figure thus, 100 — 12
and multyply 500 by 12, 500
and therof amounteth 6000, whiche yf you diuide by 100, ye quotient wyll be 60: that is the pryce of 500 for 30 myles. Then begyn the ſecond worke ſayenge: yf 30 myles coſt 60 d. how moche wyll 100 myles coſt? Set your figure thus. 30 — 60
And then multiply 100 100
by 60, wherof amounteth 6000, whiche beynge diuided by 30 wyll yelde 200. Then you may ſay that ſo many pennes ſhall coſt the caryage of 500 pounde weyghte, 100 myles after the rate of 12 d. for the 100 caryed 30 myles. S. Nowe I perceaue it alſo. M.

Queſtyō of ſo=
wynge.
Then anſwere me to this queſtion, 30 buſſhels of whete ſowed, yeldeth in one yere 360, how many wyll 80 buſ=ſhels yelde in 7 yere? I meane, ſow=
ynge

Progreſſion.

ynge euery yere of thoſe 7 ſtyll 80 buſſhelles. S. Fyrſte I ſaye, that yf 30 buſſhels yelde 360 in one yeare, thē 80 buſſhelles wyll yelde 960, in one yeare. Then for the ſeconde worke, I ſaye : yf one yeare yelde 960, then 7 yeare wyll yelde 6720, as theſe two figures do 30 ⸺ 360 1 ⸺ 960
ſhowe.   80 ⸺ 960 7 ⸺ 6270

But nowe ſyr, yf I ſet forthe 30 buſ= *Queſtion of corne.* ſhels of corne to an other man for 7 yeare, agreyng ſo, that he ſhall ſowe euery yere the hole encreaſe of ẏ corne and I at the ende of thoſe 7 yeares to haue the halfe of the hole encreaſe, I wold knowe how many buſſhels wyll there amount to my parte, ſuppoſyng the encreaſe to be after the rate of the laſt queſtion, for 30 buſſhels in 1 yere 360? M. In ſuch a queſtyon you muſt haue ſo many ſeuerall workynges, as there be yeares : as for example. In ẏͤ fyrſte yeare 30 buſſhels yelde 360, thē to knowe the yeldynge of the ſeconde yere, I muſt ſay : yf 30 yelde 360, how
P.   many

Progreſsion.

many yeldeth 360? Worke by your rule, and you ſhall fynde 4320. Then ſaye for the thyrde yere: yf 30 yelde 360, how many wyll 4320 yelde? you ſhall haue 51840, and ſo euery yeare, multyplyenge ẏ hole encreaſe by 360, and diuidynge it by 30, the encreaſe of the next yeare wyll amout, as theſe 7 figures do orderly declare,

a	b
30 —— 360	30 —— 360 360 —— 4340
c	**d**
30 —— 360 4340 —— 51840	30 —— 360 51840 —— 622080

e
30 —— 360 622080 —— 7464960
f
30 —— 360 7464960 —— 89579520
g
30 —— 360 89579520 —— 1074954240

where

Progreſſion.

where I haue ſet 7 letters for the 7 yeares, of whiche the fyrſt is ſet without arte, bycauſe that is the encreaſe whiche you do preſuppoſe, and ỹ laſte nomber of eche other doth ſhowe the encreaſe of the yeare that it ſtandeth for, which the letters doth declare, ſo that the encreaſe of the 7 yeare is 1074954240 buſſhelles: howe many quarters that is, and alſo how many weyes, you maye by Reduction ſone fynde. Now with one queſtyon more I wyll proue you. Yf 6 mowyers do mowe 45 acres in 5 dayes, how many mowyers wyll mowe 300 acres in 6 dayes? S. Yf 45 acres do requyre 6 mowyers, then 300 acres requyreth 40. Now agayne yf 5 dayes requyre 40 mowyers, then 6 dayes nedeth but 33 mowyers. M. Why do you not make mentiõ of the 2 that remayneth in the laſt diuiſion? for the laſt parte of the queſtiõ is wrought by the backer rule, where the fyrſt nomber 5, is multyplyed into the ſeconde, that is

Queſtiõ of mowynge.

P.ii.   40

Progreſſion.

40, wherof amounteth 200, whiche yf you diuide by the thyrd nomber 6, the quotient wyll be 33, as you sayd, but thē wyll there remayne 2, which can not wel be diuided into 6 partes: howe be it you may vnderſtande by y̆ 6 parte of 2, the thyrde parte of one mannes worke, whiche you muſt put to the 33: or elles you maye saye that 33 worke men wyll ende all the 300 acres in 6 dayes, saue two mens worke for one daye, or two dayes worke for one man. But suche broken nomber, called fractions, you ſhall here after more better perceaue, when J ſhall holy enſtructe you of them. But now wyll J ſhowe you of the rule of felow ſhyp or company, whiche hath sundry operations accordynge to the dyuers nōber of y̆ cōpany. This rule is somtyme without difference of tyme, and ſomtymes there is in it dyfference of tyme. Fyrſt J wyll ſpeake of that w̄out dyfference of tyme, of whiche let this be an example. Foure marchaū= tes

Rule of felow ſhyp.

Wythout tyme.

Progreſſion.

tes of one company made a bancke of monye dyuersly, for the fyrste layde in 30 li. the seconde 50 li. ẏ thyrde 60 li. and the fourth 100 li. whiche stocke they occupyed so longe, tyll it was encreaseth to 3000 li. Now I demaund of you, what shulde eche man receaue at the partynge of this monye? S. I perceaue that this rule is lyke the other, but yet there is a differēce, whiche I perceaue not. M. Then wyll I showe it to you. Fyrst by addytiō you shall brynge all the partycular summes of ẏ marchaūtes into one sūme, whiche shall be the fyrst sūme in your workynge by the golden rule : and ẏ hole sūme of the gaynes w̄ the stocke shalbe the seconde sūme. Now for the thyrde sūme you shall set the portion of eche man, one after an other, & then worke by the goldē rule, & the fourth sūme wyll showe you eche mans gaynes : as in example. The parcels of those foure marchaūtes make in one sūme 240 li. set that in the fyrst place,

Queſtiō of a bancke.

P.iii.   the

Progreſſion.

the gaynes in the ſeconde, & the fyrſte mans portion of ſtocke in the thyrde place, thus. 240 ⟋ 3000
Now multyply the 30 ⟋
ſeconde by the thyrde, and it wyll be 90000, which you ſhal diuide by 240, and there wyll appere 375 li. thus,
240 ⟋ 3000 and that is the gay=
30 ⟋ 375 nes for the fyrſt man.
Now for the ſecond man, ſet the 50 li. that he brought in the thyrde place, and worke as before, and his parte wylbe 625 li. as this figure ſhoweth.
240 ⟋ 3000 Lyke wayes for the
50 ⟋ 625 thyrde man, ſette his
monye whiche was 60 li. & his parte of gaynes wyll be 750 li. as here ap‐ pereth. 240 ⟋ 3000
And ſo for y͛ fourth 60 ⟋ 750
man, yf you ſet his ſumme, whiche is 100 li. his gaynes wyll be 1250 li. as the profe wyll declare. 240 ⟋ 3000
S. This I perceaue, 100 ⟋ 1250
but is there any way to examyn whe‐ ther I haue wel done or no? M. That muſte

Progreſſion.

muſt you do by one commyne profe, which ſerueth to the golden rule, and all other enſewynge of the ſame. And that is this : Chaunge the ſtandyn‑ ges of the nombers, and ſet the thyrd in the fyrſt place, the fourth in the ſe‑ conde place, and the fyrſt in the thyrd place, and then worke by the golden rule, and yf you haue done well, the fourth nomber now wyll be the ſame, that was the ſeconde before. As for ex ample, J wyll take the laſte worke, whiche was this : 240 — 3000 which to exampne 100 — 1250

Profe.

100 — 1250 J alter as J ſaid thus.
240 —   Now yf J multyplye y̅ secõd nõber by y̅ thyrd, & diuide that, that amounteth by the fyrſt, then wyl the fourthe nomber be 3000, whiche was the ſecond before, as you ſe here.
100 — 1250 which is a token, that
240 — 3000 J haue well done.
But as in a ſyngle rule one profe thus is ſufficient : ſo in a rule where many operations be, you muſt turne

P.iiii.   euery

Progreſsion.

euery of them, as I haue done with this one. S. Then for the profe of the fyrſt worke of this rule I ſhuld turne the nombers thus.   30 ⇗ 375
And for ỹ ſecond thus.   240 ⇙

50 ⇗ 625   And for the thyrde,
240 ⇙   thus. 60 ⇗ 750
And in eche of them, 240 ⇙
yf the workyng were true, the fourth nomber wyll be ſtyll 3000. M. Well, now an other example wyll I put to you not of gaines, but of loſſe : for one reaſon ſerueth for bothe. Yf thre marchauntes in one ſhyppe, and of one felowſhyp, had bought marchaũdyſe, ſo that the fyrſte had layd out 200 li. the ſeconde 300 li. and the thyrde 500 li. And it chaunced by tempeſt ỹ they dyd caſt ouer borde into the ſee marchaundyſe of ỹ valewe of 100 li. how moch ſhuld eche mã bere in this loſſe? S. yf I ſhall do in this as you dyd in the other queſtyõ, then muſt I ioyne theyr thre portiõs together, 200, 300, & 500, whiche maketh 1000 : then ſay

Queſtion of loſſe.

I

Progreſſion.

I, yf 1000 leſe 100, then ſhal 200 leſe 20, and 300 ſhall leſe 30, and 500 ſhal leſe 50, as by theſe .iiii. figures it doth appeare playne.

```
1000 ―― 100    1000 ―― 100
 200 ―― 20      300 ―― 30
        1000 ―― 100
         500 ―― 50
```

M. Thus you perceaue the vſe of the rule without tyme. And that you may as well perceaue the ſame with diuerſite of tyme, I propoſe this example. *The rule of fellowſhype with tyme.* iiii. marchaūtes made a cōmen ſtock, which at the yeares ende was encreaſed to 35145 li. Now to knowe what ſhalbe eche mānes porcyō of gaynes, you muſte knowe eche mans ſtocke & tyme of continuaūce. The fyrſte man *Queſtiō of a bancke.* of theſe .iiii. layde in 669 li. which he dyd take from the ſtocke agayne at ẏ ende of 10 monethes. The ſecond mā layd in 810 li. for 8 monethes. The thyrd layde in 900 li. for 7 monethes. And the fourth layde in 1040 li. for 12 monethes. This queſtion ſhal you

P.v.   exa=

Progreſſion.

exampyn as you dyd the other before, ſauynge, that where as in the thyrde place of the figure, you ſet eche mans ſume alone, here you ſhal ſet the ſame beyng multiplyed by ẙ nõber of theyr tyme, and lyke wayes in ẙ fyrſt place of the figure you ſhall ſet theyr hole ſummes ſo multiplyed by theyr tyme, and added into one ſumme, as thus. The fyrſt mans ſumme is 669 li. whi= che I multyplye by 10 (that was the nomber of his tyme) and it maketh 6690. The ſeconde mans ſume 810 li. multiplied by 8 (which was his time) maketh 6480. The thyrde mans ſume 900 li, multiplyed by 7 (for that was his tyme) yeldeth 6300. The fourthe mans ſume was 1040 li. & his tyme 12, multiply the one by the other, and it wyll be 12480. Theſe .iiii. ſummes thus multiplyed by theyr tyme, muſte be ſet orderly in the thyrd place of the figure, and in the fyrſte place muſt be ſet the hole ſumme of all .iiii. whiche is 31950. Now to ende the queſtion, I ſay fyrſte, yf 31950 dyd get 35145,
what

Progreſſion.

what dyd 6690 get? Anſwer 7359 li. as by this figure appereth.

        a
31950   35145   Lyke wayes the ſe=
6690   7359   cond mā had to his parte 7128 li. The thyzd muſt haue 6930 l. And ẏ fourth man ſhall haue for his part 13728 li. as theſe thre figures do parcelly declare.

        b   c
31950   35145   31950   35145
6480   7128   6300   6930

        d
31950   35145
12480   13728

S. This I lyke very well: but what profe is there of this worke? M. The ſame that I taughte you for ẏ other. How be it there is vſed bothe for this worke and the other alſo this maner of profe, to adde all the porcyons together, and yf they agree to the hole ſumme, then ſemeth it well done: but this is no ſure rule. S. Yet wyll I proue it in this example. The 4 pcels are theſe. whiche yf I adde together,

Profe.

    7359
    7128
    6930
   13728

there

Progreſsion.

there wyll amount 35145, & that was the hole summe, ſo is this rule true here. M. And ſo wyll it be ſtyll, when the worke is truely done. But yf you lyſt to ſe it proued falſe, take 10000 li. frō the fourth man, and put it to any of the other 3, and then be you ſure ẏ you haue not done well, and yet wyll that profe allowe it, for the addition wyll ſtyll be all one. S. It muſt nedes be ſo : but what haue I now to learne? M. There are many other excellent partes behynde, of which I wyll not as now make mencion, bycauſe that without the knowlege of fractions, they can not duely be taught, & moch leſſe vnderſtande. Therfore wyll I propoſe to you .ii. or .iii. queſtions more, wherby you maye practiſe the better the feate of the rule of felow=shyppe, and ſo make an ende for this tyme. But this maye not be forgot=ten, that in all ſuche queſtions, yf the monye be of diuers kyndes, you muſt by reduction, brynge it to one kynde,

that

Progreſſion.

that is to ſaye, to the leaſt valowre, ẏ is named in the queſtion. And lyke wayes ſhall you do, yf the tyme be of dyuers kyndes: as ſome yeares, ſome monethes, wekes, & dayes: you ſhall make all monethes, wekes, or dayes, accordynge as the leaſt name of tyme in the queſtyon is. As for example, fyrſt in dyuerſite of monye: iii. com= panyons boughte 2000 ſhepe, & payd for them 241 li. 13 s̷. 4 d. of whiche ſumme one payd 101 li. 10 s̷. The ſecōd payde 82 li. 17 s̷. 10 d. And the thyrd payde 57 li. 5 s̷. 6 d. How many ſhepe muſt eche of them haue? Anſwere: The fyrſt ſhall haue 840. The ſecōd 686. And the thyrde 474. And that muſt you worke thus: Fyrſte conſyde= rynge that your monye is of dyuers denominatiōs, you ſhall (by reductiō) brynge it all into the ſmaleſt denomi= nation, whiche is in it, that is to ſaye pennes, and ſo wyll the totall ſumme be 58000 d. Now yf you turne eche mans monye into pēnes alſo, the fyrſt mans

Queſtiō of Shepe.

Progreſſion.

mans ſūme wyll be 24360 ð. The ſe=
conde mans ſūme 19894 ð. And the
thyrde mans monye wyll be 13746 ð.
Now to knowe how many ſhepe euery
man ſhall haue, ſet the hole ſumme of
monye (that is 58000 ð.) in the fyrſte
place : and in the ſeconde place ſet the
nomber of ſhepe, and then orderly in
the thyrd place ſet eche mans monye :
and then multiplieng the thyrde & ſe=
conde ſūmes together, and dyuydyng
that ẏ amounteth, by the fyrſte, there
wyll appere the nomber of ſhepe that
eche man ought to haue, as theſe thre
figures do ſhowe.

$$
\begin{array}{cc}
58000 \overset{a}{\diagdown} 2000 & 58000 \overset{b}{\diagdown} 2000 \\
24360 \diagup 840 & 19894 \diagup 686
\end{array}
$$

$$
58000 \overset{c}{\diagdown} 2000
$$
$$
13746 \diagup 474
$$

S. Why do you ſet the monye in the
fyrſte place, ſeyng in the queſtion you
ſay 2000 ſhepe coſt 58000 ð. and not
thus 58000 ð. coſte 2000 ſhepe? M.
And

Progreſſion.

And you remembꝛe, I taught you at the begynnynge of this golden rule, that the fyꝛſt and thyꝛde nōber muſte be of one name, and of lyke thynges, and euer moꝛe the nōber that the que ſtyon is aſked of, muſte be ſet in the thyꝛde place. Nowe is the queſtyon playnly this. Yf 4 men bought 2000 foꝛ 58000 ð. how many ſhall eche mā haue. But ſeyng in this queſtiō there ought moꝛe reſpecte to be had to the ſūme of monye, then to the ſumme of the perſons (foꝛ in the ſummes of mo= nye is there pꝛopoꝛtion towarde the ſhepe, and not in the nomber of per= ſons) therfoꝛe muſt we turne the que= ſtyon, thus : Yf 58000 ð. boughte 2000 ſhepe, how many dyd 24360 ð. bye? Agayne, how many dyd 19894 ð. bye? and how many bought 13746 ð? S. I perceaue it reaſonable, & ſo ſhall I do in all lyke queſtyons. M. Euen ſo. But foꝛ eaſenes of ẏ woꝛke, marke this : when ſo euer the fyꝛſte & ſecōde nomber hath cyphers in theyꝛ fyꝛſte
places,

Progreſſion.

places, you may bothe in the multi=
plication and in the diuiſion leue out
thoſe cyphers, ſo that you leue out
lyke many out of bothe ſummes : as
in this queſtyon, ẙ fyrſte nōber 58000
hath thre cyphers, and ſo hath the ſe=
conde, that is 2000, therfore caſte a=
waye theyr cyphers, and ſo wyll the
fyrſt nomber be 58, and the ſeconde 2,
ſet them in theyr places, and worke
accordynge to the rule, and you ſhall
perceaue that it wyll be all one, ſauyng
that this is ẙ ſhorter and eaſier way,
as theſe thre figures do ſhowe.

$$58 \xrightarrow{a} 2$$
$$24360 \qquad 840$$

$$58 \xrightarrow{b} 2$$
$$19894 \qquad 686$$

$$58 \xrightarrow{c} 2$$
$$13746 \qquad 474$$

And this you ſe, is bothe the eaſyer
and alſo the more certayne waye, to
knowe the anſwere to this queſtyon.
S. Truth it is as you ſay : But ſyr, me
ſemeth I myghte aſke a farther que=
ſtyon

Progreſſion.

ſtyon here, not only how many ſhepe eche man ſhulde haue, but alſo what euery ſhepe coſte. M. That queſtyon doth not onely belonge to this rule, but maye alſo be dyſcuſſed by Dyuiſyon, eſpecyally yf the queſtyons nomber be one onely: as thus. Diuide the totall ſũme 58000 d. by 2000 (other 58 by 2, omyttynge the cyphers) and the quotient wyll be 29 d. that is 2 s. 5 d. how be it, by this rule you maye do it, and beſt, when the nomber of y̑ quotient doth excede 1: as yf I ſhulde aſke this queſtion, 2000 ſhepe coſte 58000 d. how moche dyd 20 coſte? Then ſhall I ſet my figure thus.

2000 ⎯ 58000 And doynge after
  20 ⎯   the rule there wyll

amounte 580 d. (that is 2 li. 8 s. 4 d.) the pryce of one ſhore. But yf you wyl vſe that eaſy waye, that I dyd teache you, you may chaunge the fyrſte and ſeconde nomber thus.   2 ⎯ 58
Yet now one queſtyon   20 ⎯
moze wyll I moue (that you may per

                                    D   ceaue

Progreſſion.

Queſtion. ceaue the vſe of all other lyke) and ſo make an ende. There is in a cathe=drall churche 20 cānones, and 30 vy=cars, thoſe may ſpende by yeare 2600 li. but euery cānone muſt haue to his parte 5 tymes ſo moch as euery vy=care hath: how moche is euery mans poztion, ſay you? S. I praye you make the aunſwere your ſelfe, ſo ſhall I per ceaue beſt the meanes to anſwere to ſuche other lyke. M. In this queſtiō you muſt do as in thoſe that haue di=uerſitie of tyme, for here is diuerſitie of poztions. Therfore ſhall you mul=typlye the nomber of the perſons by theyr dyfference of poztion (as you dyd in the other by tyme) then muſt you multiplye the 20 (whiche is the nōber of cānones) by 5 (for that is the nōber of theyr poztiō) ſo wyl it be 100: then 30 (that is the nōber of vicares) by 1, (that is the nomber of theyr poz tion) and it wyll be 30: put thoſe two ſummes together, and they make 130, they ſaye thus: Yf 130 ſpende 2600 li.

what

Progression.

what maye 100 spende? The rule showeth 2000 li. Agayne for the vycares: Yf 130 spende 2600 li. what maye 30 spende? Answer, 600 li. as these figures showe.

```
130 ⸺ 2600    130 ⸺ 2600
100 ⸺ 2000     30 ⸺  600
```

But yf every canone shulde haue so often tymes 4 li. as the vycare shuld haue 3 li. then shulde I multiply 20 by 4 (that were 80) and 30 by 3 (that were 90) and then bothe were 170. Then shulde the figures be set thus.

```
         li. s. d.              li. s. d.
170 ⸺ 2600         170 ⸺ 2600
 80 ⸺ 1232,10,7     90 ⸺ 1376,9,5
```

But this sorte is to harde for you by reason of the fractiōs, therfore I wyl let it reste to that place. And by this rule you se what the 20 cānons maye spende, whiche sūme yf you diuide by 20, you shall se eche cānons portion, and so of y̆ vicars, yf you diuide theyr sūme by 30, the quotient wyll declare every vicars portion.

D.ii. The

Accomptynge

⁋ The seconde dialoge
of accomptynge by
counters.
Mayster

Nowe that you haue learned the commen kyndes of Arithmetyke with the penne, you shall se the same arte in coũters: whiche feate doth not only serue for them that can not write and rede, but also for them that can do bothe, but haue not at some tymes theyr penne or tables redye with them. This sorte is in two fourmes cõmenly. The one by lynes, and the other without lynes: in that ẏ hath lynes, the lynes do stande for the order of places: and in ẏ that hath no lynes, there muste be sette in theyr stede so many counters as shall nede, for eche lyne one, and they shall supplye the stede of the lynes. S. By examples I shuld better pceaue your meanynge. M. For example of the ly=
nes

by counters.

nes : Lo here you ſe .vi. lynes whiche
ſtande foꝛ ſyxe —————100000—————
places, ſo that —————10000—————
the nethermoſt ✳—1000—————
ſtandeth foꝛ ẙ —————100—————
fyꝛſt place, and —————10—————
the nexte aboue —————1—————
it, foꝛ the ſecond : and ſo vpward tyll
you come to the hygheſt, which is the
ſyxte lyne, and ſtandeth foꝛ the ſyxte
place. Now what is the valewe of e=
uery place oꝛ lyne, you may perceaue
by the figures whiche I haue ſet on
them, whiche is accoꝛdynge as you
learned befoꝛe in the Numeration of   Numera-
figures by the penne : foꝛ the fyꝛſte   tion.
place is the place of vnities oꝛ ones,
and euery counter ſet in that lyne be=
tokeneth but one : ⁌ the ſeconde lyne
is the place of 10, foꝛ euery counter
there, ſtandeth foꝛ 10. The thyꝛd lyne
the place of hundredes : the fourth of
thouſandes : ⁌ ſo foꝛth. S. Syꝛ I do
perceaue that the ſame oꝛder is here
of lynes, as was in the other figures

             M.iii.   by

Accomptynge

by places, so that you shall not nede longer to stande about Numeration, excepte there be any other difference. M. Yf you do vnderstãde it, then how wyll you set 1543? S. Thus, as I suppose. M. You haue set ẏ places truely, but your figures be not mete for this vse: for the metest figure in this behalfe, is the figure of a coũter round, as you se here, where I haue expres=
sed that same summe.
S. So that you haue not one figure for 2, nor 3, nor 4, and so forth, but as many digettes as you haue, you set in the lowest lyne: and for euery 10 you set one in the second line: and so of other. But I know not by what reason you set that one counter for 500 betwene two lynes. M. you shall remember this, that when so euer you nede to set downe 5, 50, or 500, or 5000, or so forth any other nomber, whose numerator is 5,

by counters.

is 5, you shall set one counter for it, in the nexte space aboue the lyne that it hath his denomination of, as in this example of that 500, bycause the numerator is 5, it must be set in a voyd space: and bycause the denominator is hundred, I knowe that his place is the voyde space nexte aboue hundredes, that is to say, aboue the thyrd lyne. And farther you shall marke, that in all workynge by this sorte, yf you shall sette downe any summe betwene 4 and 10, for the fyrste parte of that nomber you shall set downe 5, & then so many counters more, as there reste nōbers aboue 5. And this is true bothe of digettes and articles. And for example I wyll set downe this sūme 287965, which sūme yf you marke well, you nede none other exāples for to lerne the numeration of

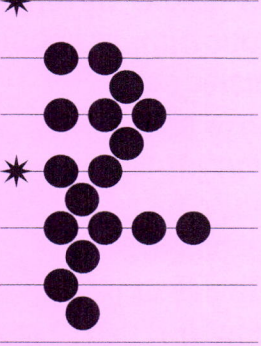

this

Accomptynge

this forme. But this shal you marke, that as you dyd in the other kynde of arithmetike, set a pricke in the places of thousādes, in this worke you shall sette a starre, as you se here. S. Then I perceaue numeration, but I praye you, howe shall I do in this arte to adde two summes or more together? M. The easyest way in this arte is, to adde but 2 sūmes at ones together: how be it you maye adde more, as I wyll tell you anone. Therfore when you wyll adde two sūmes, you shall fyrst set downe one of them, it forseth not whiche, & then by it drawe a lyne crosse the other lynes. And afterward set downe the other sūme, so that that lyne may be betwene them, as yf you wolde adde 2659 to 8342, you must set your sūmes as you se here. And then yf you lyst, you

Subtraction

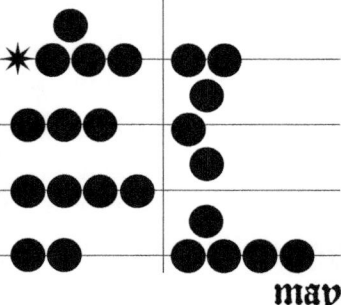

may

by counters.

may adde the one to the other in the
same place, oz els you may adde them
both together in a newe place : which
waye, bycause it is moste playnest, I
wyll showe you fyzst. Therfoze wyl I
begynne at the vnites, whiche in the
fyzst sūme is but 2, & in y̒ second sūme
9, that maketh 11, those do I take
vp, and foz them I set 11 in the newe
roume, thus.

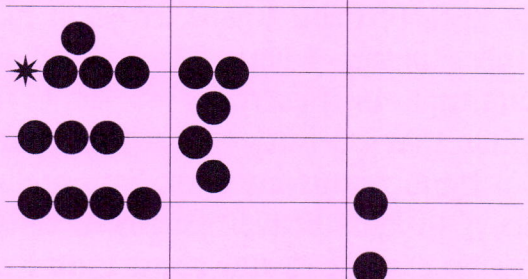

Then do I take vp all y̒ articles vn=
der a hundred, which in the fyzst sūme
are 40, and in the seconde summe 50,
that maketh 90 : oz you may saye bet=
ter, that in the fyzste summe there are
4 articles of 10, and in the seconde
summe 5, whiche make 9, but then
take hede that you sette them in theyz
M.v.   ryght

Accomptynge

ryght lynes, as you ſe here.

Where I haue taken awaye 40 frō the fyrſte ſume, and 50 from ẙ ſecond, and in theyr ſtede I haue ſet 90 in the thyrde, whiche I haue ſet playnely ẙ you myght well perceaue it : how be it ſeynge that 90 with the 10 that was in ẙ thyrd roume all redy, doth make 100, I myghte better for thoſe 6 coū= ters ſet 1 in the thyrde lyne, thus :

For it is all one ſumme as you may ſe, but it is beſte, neuer to ſet 5 coū= ters in any line, for that may be done with 1 coū ter in a hygher place. S. I iudge that good reaſō, for many are vnnedefull, where one wyll ſerue. M. Well, then
wyll

by counters.

wyll I adde forth of hundredes : I fynde 3 in the fyrste summe, and 6 in the seconde, whiche make 900, then do I take vp & set in the thyrd roume where is one hundred all redy, to whiche I put 900, and it wyll be 1000, therfore I set one coūter in the fourth lyne for them all, as you se here.

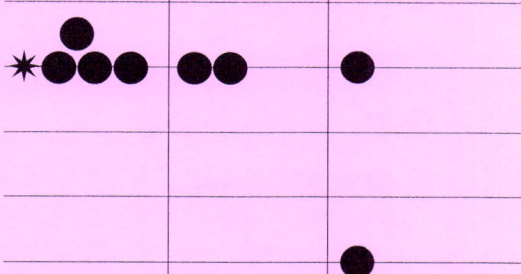

Then adde I ẙ thousandes together, whiche in the fyrst summe are 8000, & in ẙ second 2000, that maketh 10000 : them do I take vp frō those two places, and for them I set one counter in the fyxte lyne, and then appereth as you se, to be 11001, for so many doth amount of the addition of 8342 to 2659.

S.

Accomptynge

S. Syz, this I do perceaue : but how shall I set one sume to an other, not chaungynge them to a thyrde place? M. Marke well how I do it : I wyll adde together 65436, and 3245, which fyrste I set downe thus.

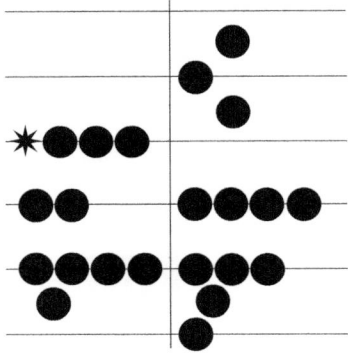

Then do I begynne with the smalest, which in the fyrst summe is 5, that do I take vp, and wold put to the other 5 in the seconde summe, sauynge that two counters can not be set in a voyd place of 5, but for them bothe I must set 1 in the seconde lyne, which is the place of 10, therfore I take vp the 5 of the fyrst sume, & the 5 of the secōde, and for them I set 1 in the secōd lyne,

as

by counters.

as you se here.

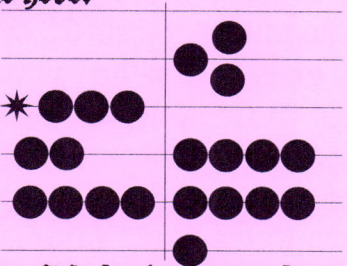

Then do I lyke wayes take vp the 4 counters of the fyrste sume & seconde lyne (which make 40) and adde them to the 4 counters of the same lyne, in the seconde sume, and it maketh 80, But as I sayde I maye not conueniently set aboue 4 coūters in one lyne, therfore to those 4 that I toke vp in the fyrst sume, I take one also of the seconde sume, and then haue I taken vp 50, for whiche 5 counters I sette downe one in the space ouer y̆ second lyne, as here dothe appere.

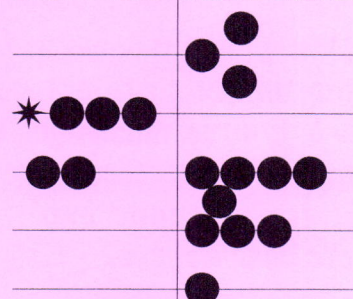

Accomptynge

and then is there 80, as well ⱳ thoſe 4 counters, as yf I had ſet downe ẏ other 4 alſo. Now do I take the 200 in the fyrſte ſume, and adde them to the 400 in the ſeconde ſumme, and it maketh 600, therfore I take vp the 2 counters in the fyrſte ſumme, and 3 of them in the ſeconde ſumme, and for them 5 I ſet 1 in ẏ ſpace aboue, thus.

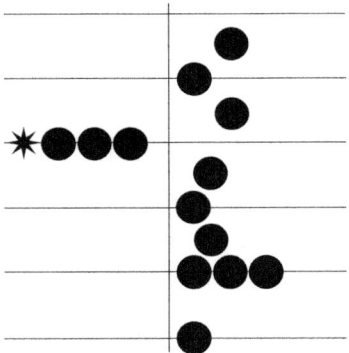

Then I take ẏ 3000 in ẏ fyrſte ſume, vnto whiche there are none in the ſe= cond ſumme agreynge, therfore I do onely remoue thoſe 3 counters from the fyrſte ſumme into the ſeconde, as here doth appere.

And

by counters.

And so you see
the hole sūme,
that amoūteth
of the addytiō
of 65436 with
3245, to be 6868
And yf you ha=
ue marked the=
se two exāples

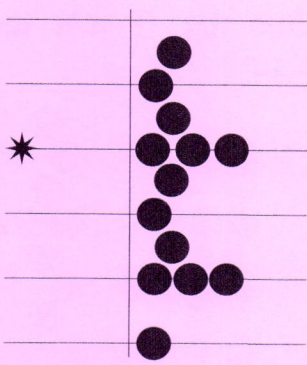

well, you nede no farther enstructiō in
Addition of 2 only summes : but yf
you haue moze then two summes to
adde, you may adde them thus. Fyzst
adde two of them, and then adde the
thyzde, and ẏ fourth, oz moze yf there
be so many : as yf I wolde adde 2679
with 4286 and 1391. Fyzst I adde
the two fyzste summes thus.

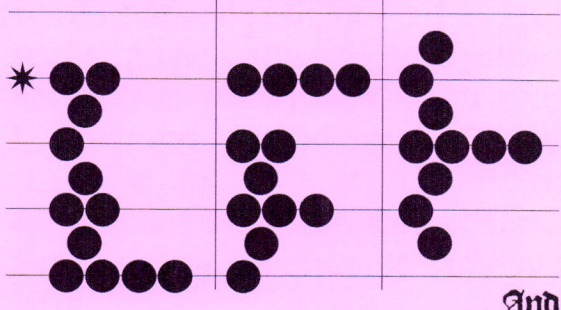

And

Accomptynge

And then I adde the thyrde there to thus.

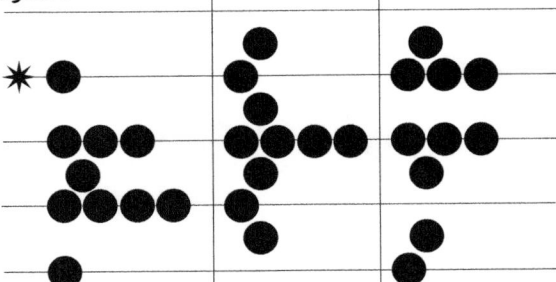

And so of moꝛe yf you haue them. S. Nowe I thynke beste that you passe foꝛth to Subtraction, except there be any wayes to exampyn this maner of Addition, then I thynke that were good to be knowen nexte. M. There is the same proofe here that is in the other Addition by the penne, I meane Subtraction, foꝛ that onely is a sure waye : but consyderynge that Subtraction must be fyꝛste knowen, I wyl fyꝛste teache you the arte of Subtraction, and that by this example : I wolde subtracte 2892 out of 8746. These summes must I set downe, as I dyd in Addition : but here it is best to set

by counters.

to set the lesser nōber fyrste, thus.

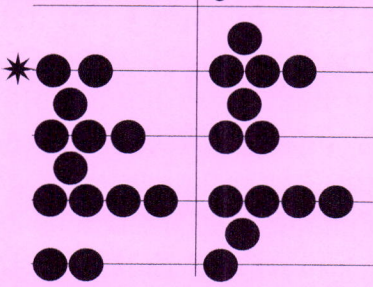

Then shall I begynne to subtracte the greatest nombres fyrste (contrary to the vse of the penne) ẏ is the thou=sandes in this exāple : therfore I fynd amongest the thousandes 2, for which I withdrawe so many frō the seconde summe (where are 8) and so remay=neth there 6, as this exāple showeth.

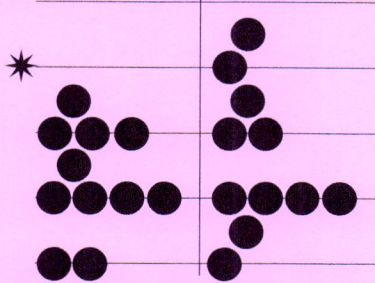

Then do I lyke wayes with the hun=dredes, of whiche in the fyrste summe
R. I

Accomptynge

I fynde 8, and is the seconde summe but 7, out of whiche I can not take 8, therfore thus muste I do : I muste loke how moche my summe dyffereth from 10, whiche I fynde here to be 2, then must I bate for my sume of 800, one thousande, and set downe the ex=cesse of hundredes, that is to saye 2, for so moche 100 is more then I shuld take vp. Therfore frō the fyrste sūme I take that 800, and from the second sume where are 6000, I take vp one thousande, and leue 5000 : but then set I downe the 200 vnto the 700 ẏ are there all redye, and make them 900 thus.

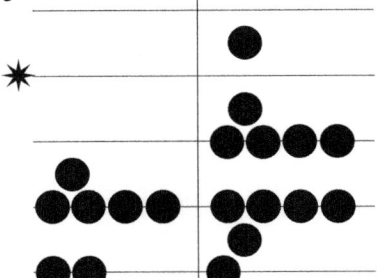

Then come I to the articles of tēnes, where in the fyrste sūme I fynde 90, and

by counters.

and in the seconde summe but only 40 : Now consyderyng that 90 can not be bated from 40, I loke how moche y̆ 90 doth dyffer from the nexte summe aboue it, that is 100 (or elles whiche is all to one effecte, I loke how moch 9 doth dyffer frõ 10) & I fynd it to be 1, then in the stede of that 90, I do take from the second summe 100 : but consyderynge that it is 10 to moche, I set downe 1 in y̆ nexte lyne beneth for it, as you se here. Sauynge that here I haue set one counter in y̆ space in stede of 5 in y̆ nexte lyne. And thus haue I subtracted all saue two, whiche I must bate from the 6 in the seconde summe, and there wyll remayne 4, thus. So y̆ yf I subtacte 2892 frõ 8746, the remayner wyll be 5854,

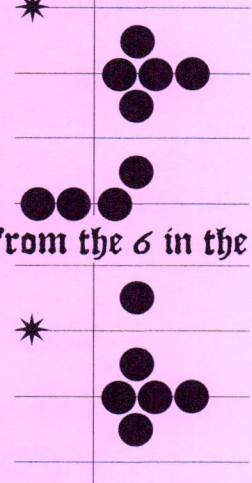

R.ii.   And

Accomptynge

And that this is truely wrought, you maye proue by Addition: for yf you adde to this remayner the same sume that you dyd subtracte, then wyll the formar sume 8746 amount agayne. S. That wyll I proue: and fyrst I set the sume that was subtracted, which was 2892, & then the remayner 5854, thus.

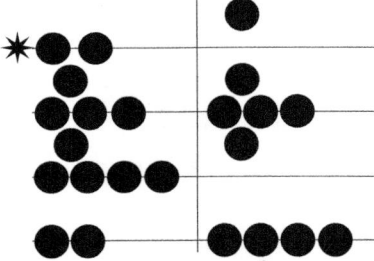

Then do I adde fyrst ye 2 to 4, whi= che maketh 6, so take I vp 5 of those counters, and in theyr stede I sette 1 in the space, as here appereth.

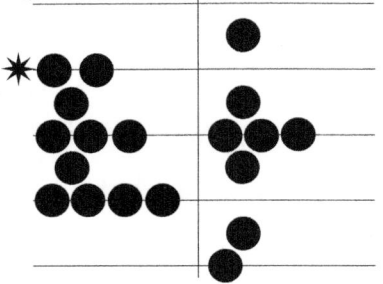

Then

by counters.

Then do I adde the 90 nexte aboue to the 50, and it maketh 140, therfore I take vp those 6 counters, and for them I sette 1 to the hundredes in ẙ thyrde lyne, & 4 in ẙ second lyne, thus.

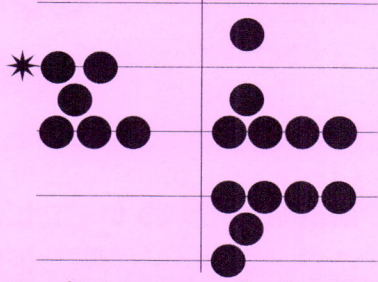

Then do I come to the hundredes, of whiche I fynde 8 in the fyrst summe, and 9 in ẙ second, that maketh 1700, therfore I take vp those 9 counters, and in theyr stede I sette 1 in the .iiii. lyne, and 1 in the space nexte beneth, and 2 in the thyrde lyne, as you se here. Then is there lefte in the fyrste summe but only 2000, whi= che I shall take vp from thence, and set

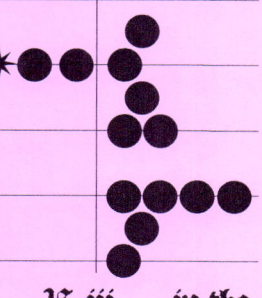

R.iii. in the

Accomptynge

in the same lyne in y̨ second sūme, to y̨ one y̨ is there all redy : & then wyll the hole sūme appere (as you may wel se) to be 8746, which was y̨ fyrst grosse summe, & therfore I do per= ceaue, that I hadde well subtracted be=

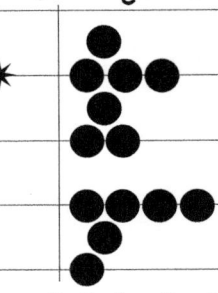

fore. And thus you may se how Sub traction maye be tryed by Addition. S. I perceaue the same order here w̃ coūters, y̨ I lerned before in figures. M. Then let me se howe can you trye Addition by Subtraction. S. Fyrste I wyl set forth this exāple of Additiō where I haue added 2189 to 4988, & the hole sūme appereth to be 7177.

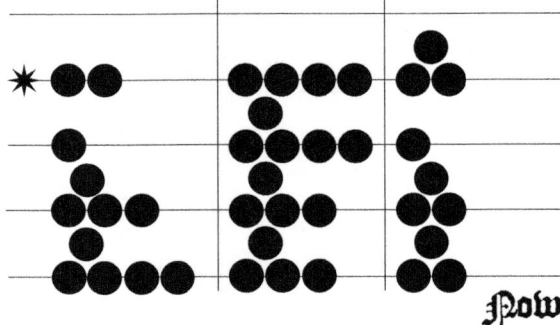

Now

by counters.

Nowe to trye whether that sūme be well added oʒ no, I wyll subtract one of the fyʒst two sūmes from the thyʒd, and yf I haue well done ẏ remayner wyll be lyke that other sūme. As foʒ example: I wyll subtracte the fyʒste summe from the thyʒde, whiche I set thus in theyʒ oʒder.

Then do I subtract 2000 of the fyʒste summe frõ ẏ second sūme, and then remayneth there 5000 thus. Then in the thyʒd lyne, I subtract ẏ 100 of the fyʒste summe, frõ the se= cond sūme, where is onely 100 also, & then in ẏ thyʒde lyne resteth nothyng. Then in the sec= onde lyne with his space ouer hym, I fynde 80, which I shuld subtract

R.iiii. from

Accomptynge

from the other ſume, then ſeyng there are but only 70 I muſt take it out of ſome hygher ſumme, whiche is here only 5000, therfoꝛe I take vp 5000, and ſeyng that it is to moch by 4920, I ſette downe ſo many in the ſeconde roume, whiche with the 70 beynge there all redy do make 4990, & then the ſummes doth ſtande thus.

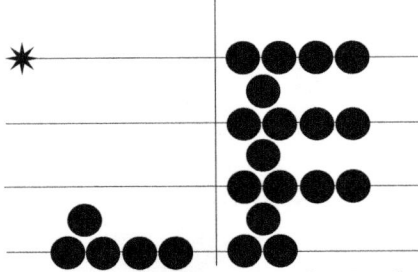

Yet remayneth there in the fyꝛſt ſume 9, to be bated from the ſecond ſumme, where in that place of vnities dothe appere only 7, then I muſte bate a hygher ſumme, that is to ſaye 10, but ſeynge that 10 is moꝛe then 9 (which I ſhulde abate) by 1, therfoꝛe ſhall I take vp one counter from the ſeconde lyne, & ſet downe the ſame in the fyꝛſt

by counters.

oz loweſt lyne, as you ſe here.
And ſo haue J en=
ded this wozke, ⁊
the ſume appereth
to be ẏ ſame, whi=
che was ẏ ſeconde
ſumme of my addi
tion, and therefoze

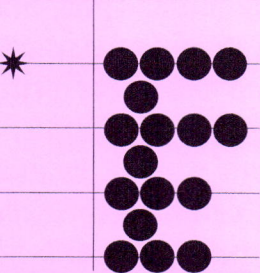

J perceaue, J haue wel done. M. To
ſtande longer about this, it is but fo=
lye : excepte that this you maye alſo
vnderſtande, that many do begynne
to ſubtracte with counters, not at the
hygheſt ſume, as J haue taught you,
but at the nethermoſte, as they do vſe
to adde : and when the ſumme to be a=
batyd, in any lyne appeareth greater
then the other, then do they bozowe

one of the nexte
hygher roume,
as foz example :
yf they ſhuld a=
bate 1846 from
2378, they ſet ẏ
ſummes thus.

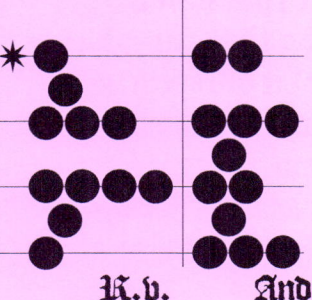

R.v.   And

Accomptynge

And fyrst they take 6 whiche is in the lower lyne, and his space from 8 in the same roumes, in y̆ second sūme, and yet there remayneth 2 counters in the lowest lyne. Then in the second lyne must 4 be subtracte from 7, and so remayneth there 3. Then 8 in the thyrde lyne and his space, from 3 of the second summe can not be, therfore do they bate it from a hygher roume, that is, from 1000, and bycause that 1000 is to moch by 200, therfore must I sette downe 200 in the thyrde lyne, after I haue taken vp 1000 from the fourth lyne : then is there yet 1000 in the fourth lyne of the fyrst sūme, whi=che yf I withdrawe from the seconde summe, then doth all y̆ figures stande in this order.

So that (as you se) it differeth not greatly whether you begynne sub=tractiō at the hy=gher lynes, or at

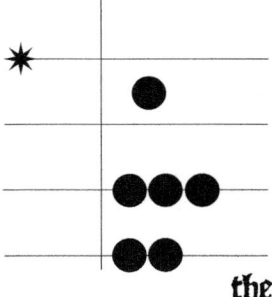

the

by counters.

the lower. How be it, as some menne lyke the one waye beste, so some lyke the other: therfore you now knowyng bothe, may vse whiche you lyst. But now touchynge Multiplicatiō: you shall set your nōbers in two roumes, as you dyd in those two other kyndes but so that the multiplier be set in the fyrste roume. Then shall you begyn with the hyghest nōbers of y̆ seconde roume, and multiply them fyrst after this sort. Take that ouermost lyne in your fyrst workynge, as yf it were the lowest lyne, settyng on it some mouable marke, as you lyste, and loke how many counters be in hym, take them vp, and for them set downe the hole multyplyer, so many tymes as you toke vp connters, reckenyng, I saye that lyne for the vnites: & when you haue so done with the hygheest nōber then come to the nexte lyne beneth, & do euen so with it, and so with y̆ next, tyll you haue done all. And yf there be any nomber in a space, then for it
                          shall

Multiplicatyon

Accomptynge

ſhall you take ẏ multiplyer 5 tymes, and then muſte you recken that lyne foꝛ the vnites whiche is nexte beneth that ſpace: oꝛ els after a ſhoꝛter way, you ſhall take only halfe the multy= plyer, but then ſhall you take the lyne nexte aboue that ſpace, foꝛ the lyne of vnites: but in ſuche woꝛkynge, yf chaūce your multyplyer be an odde nomber, ſo that you can not take the halfe of it iuſtly, then muſte you take the greater halfe, and ſet downe that, as if that were the iuſte halfe, and farther you ſhall ſet one coūter in the ſpace beneth that line, which you rec= ken foꝛ the lyne of vnities, oꝛ els only remoue foꝛward the ſame that is to be multyplyed. S. Yf you ſet foꝛth an example hereto I thynke I ſhal per= ceaue you. M. Take this exāple: I wold multiply 1542 by 365, therfoꝛe I ſet ẏ nombers thus.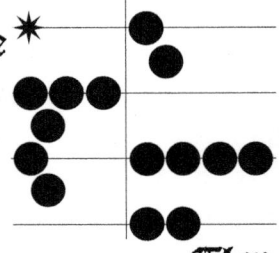

Then

by counters.

Then fyrste I begynne at the 1000 in
ye hyghest roume, as yf it were ye fyrst
place, & I take it vp, settyng downe
for it so often (that is ones) the mul=
typlyer, which is 365, thus, as you se
here: where for the one counter taken

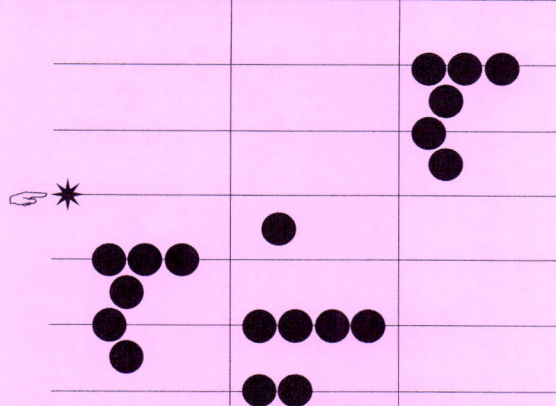

vp from the fourth lyne, I haue sette
downe other 6, whiche make ye summe
of the multyplyer, reckenynge that
fourth lyne, as yf it were the fyrste:
whiche thyng I haue marked by the
hand set at the begynnyng of ye same,
S. I perceaue this well: for in dede,
this summe that you haue set downe
is 365000, for so moche doth amount
of

Accomptynge

of 1000, multiplyed by 365. M. Well thē to go forth, in the nexte space I fynde one counter which I remoue forward but take not vp, but do (as in such case I must) set downe the greater halfe of my multiplier (sayng it is an odde nōber) which is 182, ⁊ here I do styll let that fourth place stand, as yf it were ye fyrst: as in this fourme you se,

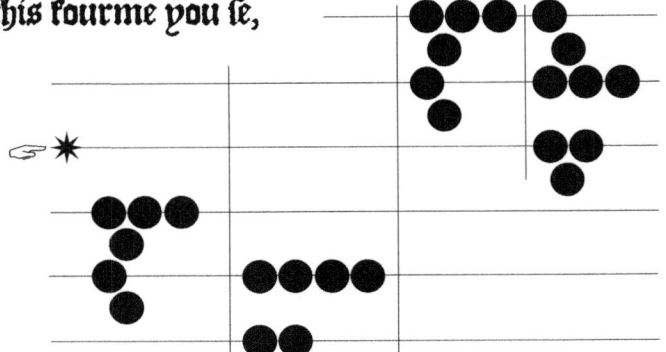

where I haue set this multiplycatiō with ye other: but for the ease of your vnderstādynge, I haue set a lytell lyne betwene them: now shulde they both in one sume stand thus.

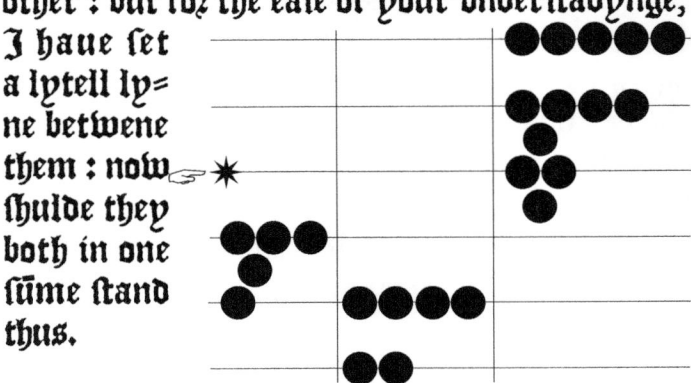

by counters.

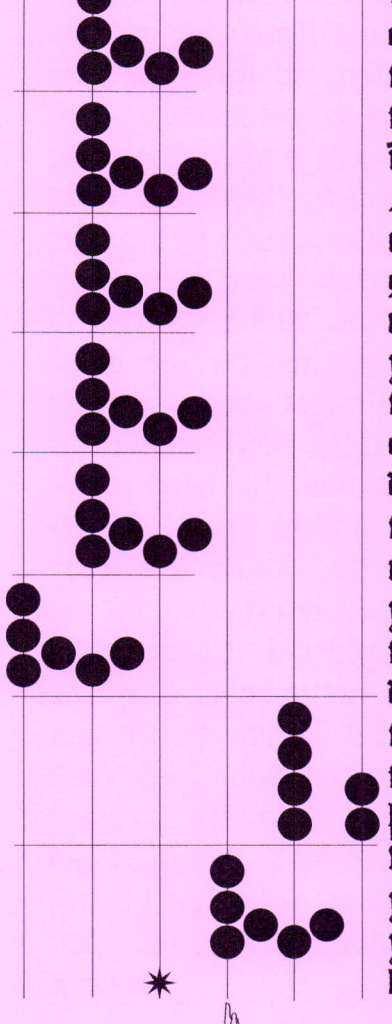

Howe be it an other fourme to multyplye suche coũters ĩ space is this: Fyrſt to remoue the kynger to the lyne nexte benethe yͤ ſpace, & then to take vp yͤ coũter, & to ſet downe yͤ multiplyer .v. tymes, as here you ſe. Which ſũmes yf you do adde together into one ſũme, you ſhal perceaue that it wyll be yͤ ſame yͤ appeareth of yͤ other workig before, ſo that bothe

Accomptynge

bothe sortes are to one entent, but as the other is more shorter, so this is playner to reason, for suche as haue had small exercyse in this arte. Not withstandynge you maye adde them in your mynde before you sette them downe, as in this exāple, you myghte haue sayde 5 tymes 300 is 1500, & 5 tymes 60 is 300, also 5 tymes 5 is 25, whiche all put together do make 1825, which you maye at one tyme set downe yf you lyste. But nowe to go forth, I must remoue the hand to the nexte counters, whiche are in the se= cond lyne, and there must I take vp those 4 counters, settynge downe for them my multiplyer 4 tymes, which thynge other I maye do at 4 tymes seuerally, or elles I may gather that hole summe in my mynde fyrste, and then set it downe : as to saye 4 tymes 300 is 1200 : 4 tymes 60 are 240 : and 4 tymes 5 make 20 : p̄ is in all 1460, p̄ shall I set downe also, as here you se. whiche

by counters.

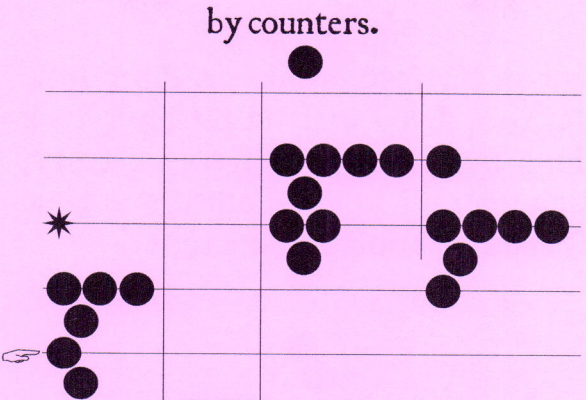

whiche yf I ioyne in one summe with the formar nombers, it wyll appeare thus.

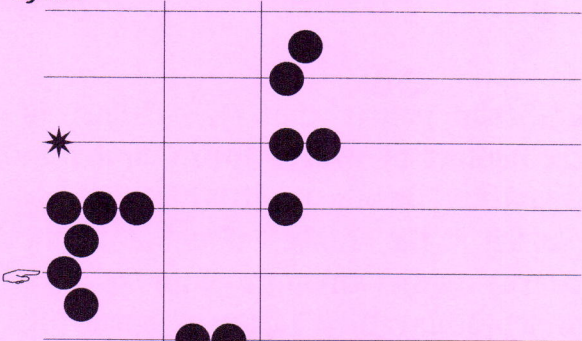

Then to ende this multiplycation, I remoue the fynger to the lowest lyne, where are onely 2, them do I take vp, and in theyr stede do I set downe twyse 365, that is 730, for which I set S. one

Accomptynge

one in the space aboue the thyrd lyne
for 500, and 2 more in the thyrd lyne
with that one that is there all redye,
and the reste in theyr order, & so haue
I ended the hole summe thus.

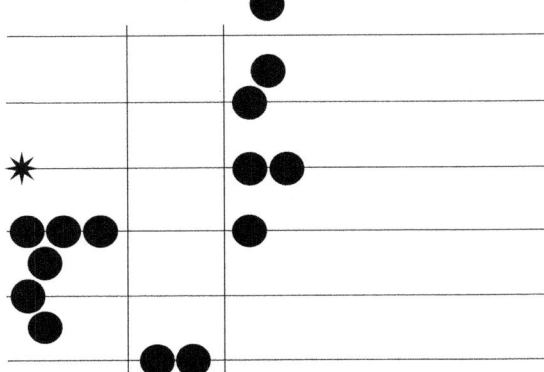

Wherby you se, that 1542 (which is
the nomber of yeares syth Chrystes in
carnation) beyng multyplyed by 365
(which is the nomber of dayes in one
yeare) dothe amounte vnto 562830,
which declareth ẏ nōber of daies sith
Chrystes incarnatiō vnto the ende of
1342 yeares. (besyde 385 dayes and 12
houres for lepe yeares) S. Now wyll
I proue by an other exāple, as this :
40 labourers (after 6 d. ẏ day for eche
man) haue wrought 28 dayes, I wold
know

by counters.

know what theyr wages doth amout vnto: In this case muste I worke doublely: fyrst I must multyplye the nomber of the labourers by ẙ wages of a man for one day, so wyll ẙ charge of one daye amount: then secondarely shall I multyply that charge of one daye, by the hole nomber of dayes, & so wyll the hole summe appeare: fyrst therfore I shall set the sumes thus.

Where in the fyrste space is the multy=plyer (ẙ is one dayes wages for one man) & in the second space is set the nomber of 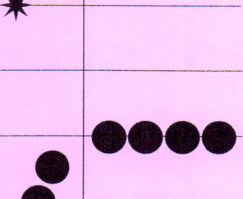 the worke men to be multyplyed: thē saye I, 6 tymes 4 (reckenynge that second lyne as the lyne of vnites) ma=keth 24, for whiche summe I shulde set 2 counters in the thyrde lyne, and 4 in the seconde, therfore do I set 2 in the thyrde lyne, and let the 4 stand styll in the seconde lyne, thus.

S.ii. So

Accomptynge

So apwereth the hole dayes wages to be 240 d. that is 20 s. Then do I multiply agayn the same summe by the nōber of dayes and fyrste I sette the nombers, thus.

Thē bycause there are couters in dyuers lynes, I shall begynne with the hyghest, and take them vp, settynge for them the multyplyer so many tymes, as I toke vp counters, y̌ is twyse, then wyll y̌ sūme stande thus.

Then come I to y̌ seconde lyne, and take vp those 4 coū ters, settynge for them the multiply= er foure tymes, so wyll the hole summe appeare thus.

So is

by counters.

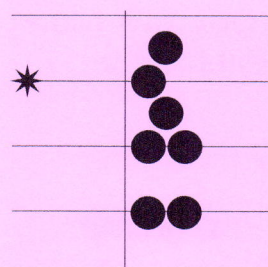

So is the hole wa{-}
ges of 40 workemē,
for 28 dayes (after
6 d̛. eche daye for a
man) 6720 d̛. that
is 560 s̛. or 28 li.

M. Now if you wold
proue Multiplycatiō, the surest way
is by Dyuision: therfore wyll J ouer
passe it tyll J haue taught you ẏ arte
of Diuision, whiche you shall worke
thus. Fyrste sette downe the Diuisor
for feare of forgettynge, and then set
the nomber that shalbe deuided, at ẏ
ryghte syde, so farre from the diuisor,
that the quotient may be set betwene
them: as for example: Yf 225 shepe
cost 45 li. what dyd euery shepe cost?
To knowe this, J shulde diuide the
hole summe, that is 45 li. by 225, but
that can not be, therfore must J fyrste
reduce that 45 li. into a lesser deno{-}
mination, as into shyllynges: then J
multiply 45 by 20, and it is 900, that
summe shall J diuide by the nōber of

S.iii. shepe.

Diuision

Accomptynge

ſhepe, whiche is 225, theſe two nom=
bers therfoze J ſette thus.

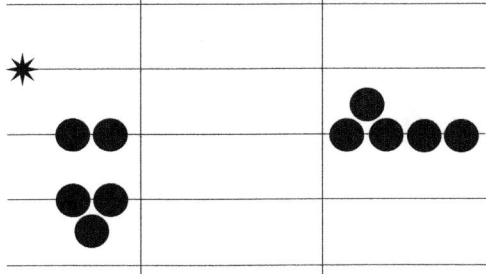

Then begynne J at the hyghest lyne
of the diuident, and ſeke how often J
may haue the diuiſoz therin, and that
maye J do 4 tymes, then ſay J, 4 ty=
mes 2 are 8, whiche yf J take from
9, there reſteth but 1, thus.

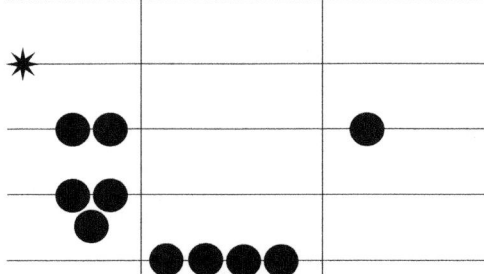

And bycauſe J founde the diuiſoz 4
tymes in the diuidente, J haue ſet (as
you ſe) 4 in the myddle roume, which
is

by counters.

is the place of the quotient: but now must I take the reste of the diuisor as often out of the remayner: therfore come I to the seconde lyne of the di=uisor, sayeng 2 foure tymes make 8, take 8 from 10, & there resteth 2, thus.

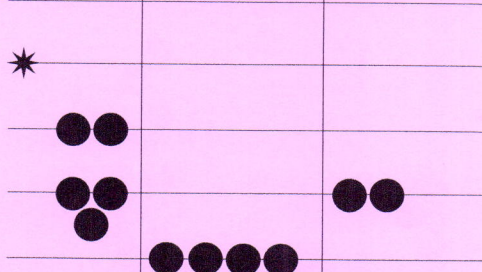

Then come I to the lowest nomber, whiche is 5, and multyply it 4 tymes, so is it 20, that take I from 20, and there remayneth nothynge, so that I se my quotient to be 4, whiche are in valewe shyllynges, for so was the di=uident: and therby I knowe, that yf 225 shepe dyd coste 45 li. euery shepe coste 4 s. S. This can I do, as you shall perceaue by this exāple: Yf 160 sowldyars do spende euery moneth 68 li. what spendeth eche man? Fyrst

S.iiii. by=

Accomptynge

bycause I can not diuide the 68 by 160, therfore I wyll turne the poūdes into pennes by multiplicatiō, so shall there be 16320 ð. Nowe muste I di=uide this sūme by the nomber of sowl oyars, therfore I set thē i order, thus.

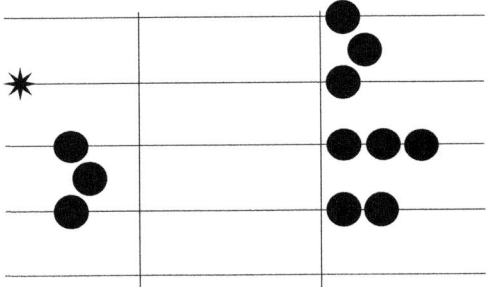

Then begyn I at the hyghest place of the diuidente, sekynge my diuisor there, whiche I fynde ones, therfore set I 1 in the nether lyne. M. Not in the nether line of the hole summe, but in the nether lyne of that worke, whi=che is the thyrde lyne. S. So standeth it with reason. M. Then thus do they stande.

Then

by counters.

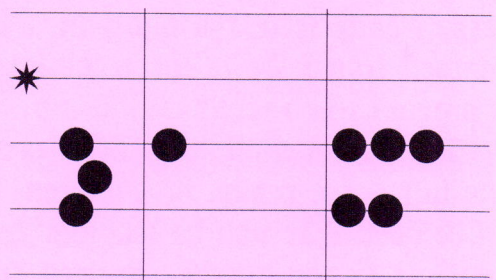

Then seke I agayne in the reste, how often I may fynde my diuisor, and I se that in the 300 I myghte fynde 100 thre tymes, but then the 60 wyll not be so often founde in 20, therfore I take 2 for my quotient: then take I 100 twyse from 300, and there resteth 100, out of whiche with the 20 (that maketh 120) I may take 60 also twyse, and then standeth the nombers thus,

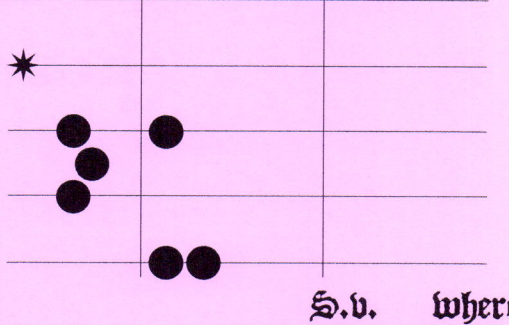

S.v.   where

Accomptynge

where I haue sette the quotient 2 in the lowest lyne: So is every sowldy= ars portion 102 d. that is 8 s. 6 d.
M. But yet bycause you shall per= ceaue iustly the reason of Diuision, it shall be good that you do set your di= uisor styll agaynst those nombres frō whiche you do take it: as by this ex= ample I wyll declare. If ẏ purchace of 200 acres of ground dyd coste 290 li. what dyd one acre coste? Fyrst wyl I turne the poundes into pennes, so wyll there be 69600 d. Then in set= tynge downe these nombers I shall do thus. Fyrst set the diuident on the ryghte hande as it oughte, and then

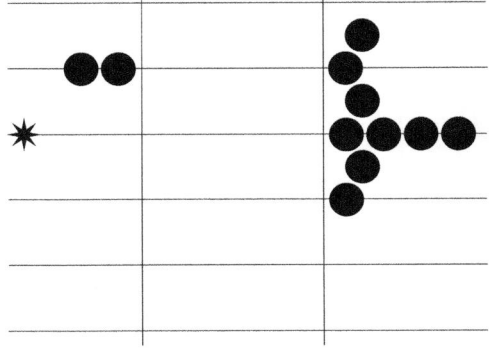

the

by counters.

the diuisor on the lefte hande agaynst those nombers, frõ which I entende to take hym fyrst as here you se, wher I haue set the diuisor two lynes hygher thē is theyr owne place. S. This is lyke the order of diuision by the penne. M. Truth you say, and nowe must I set y͞ quotient of this worke in the thyrde lyne, for that is the lyne of vnities in respecte to the diuisor in this worke. Then I seke howe often the diuisor maye be founde in the diuident, & that I fynde 3 tymes, then set I 3 in the thyrde lyne for the quotient, and take awaye that 60000 frõ the diuident, and farther I do set the diuisor one line lower, as you se here.

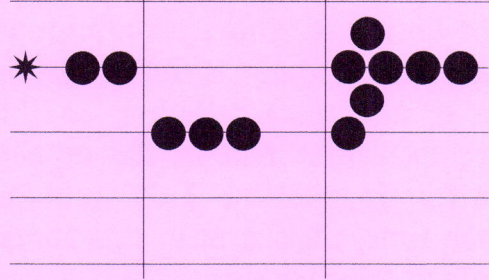

And

Accomptynge

And then seke I how often the diui=
soz wyll be taken from the nomber a=
gaynste it, whiche wyll be 4 tymes
and 1 remaynynge. S. But what yf
it chaunce that when the diuisoz is so
remoued, it can not be ones taken out
of the diuident agaynste it? M. Then
must the diuisoz be set in an other line
lower. S. So was it in diuision by
the penne, and therfoze was there a
cypher set in the quotient : but howe
shall that be noted here? M. Here ne=
deth no token, foz the lynes do repze=
sente the places : onely loke that you
set your quotient in that place which
standeth foz vnities in respecte of the
diuisoz : but now to returne to the ex=
ample, I fynde the diuisoz 4 tymes
in the diuidente, and 1 remaynynge,
foz 4 tymes 2 make 8, which I take
from 9, and there resteth 1, as this
figure sheweth : and in the myddle
space foz the quotient I set 4 in the
seconde lyne, whiche is in this wozke
the place of vnities.

Then

by counters.

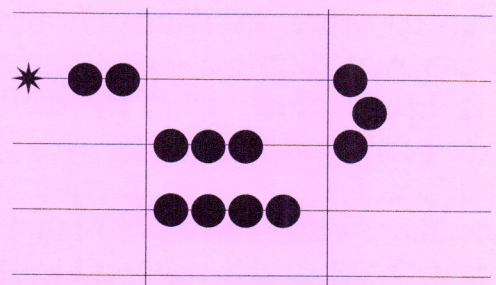

Then remoue I ẏ diuisor to the next lower line, and seke how often I may haue it in the dyuident, which I may do here 8 tymes iust, and nothynge remayne, as in this fourme,

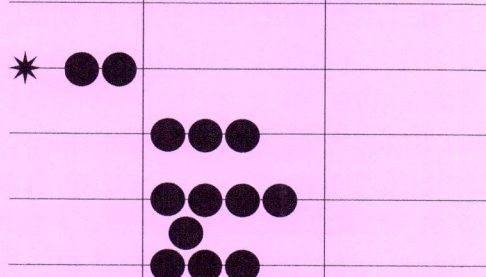

where you may se, that the hole quo=
tient is 348 d̃. that is 29 s̃. wherby
I knowe that so moche coste the pur=
chace of one aker. S. Now resteth the
profes of Multiplycatiō, and also of
Diuisiō. M. Ther best profes are eche
one

Accomptynge

one by the other, for Multyplication is proued by Diuision, and Diuision by Multiplycation, as in the worke by the penne you learned. S. Yf that be all, you shall not nede to repete a=gayne that, ỹ was sufficyētly taughte all redye: and excepte you wyll teache me any other feate, here maye you make an ende of this arte I suppose. M. So wyll I do as touchynge hole nomber, and as for broken nomber, I wyll not trouble your wytte with it, tyll you haue practised this so well, ỹ you be full perfecte, so that you nede not to doubte in any poynte that I haue taught you, and thenne maye I boldly enstructe you in ỹ arte of frac=tions or broken nōber, wherin I wyll also showe you the reasons of all that you haue nowe learned. But yet be=fore I make an ende, I wyll showe you the order of cōmen castyng, wher in are bothe pennes, shyllynges, and poundes, procedynge by no groun=ded reason, but onely by a receaued
fourme

by counters.

fourme, and that dyuersly of dyuers men : for marchautes vse one fourme, and auditors an other : But fyrste for marchauntes fourme marke this example here, in which I haue expressed this summe 168 li. 19 s̄. 11 d. So that you maye se that the lowest lyne serueth for penes, the nexte aboue for shyl= 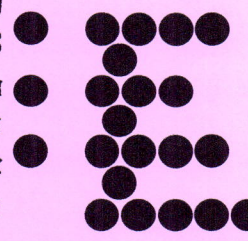 lynges, the thyrde for poundes, and the fourth for scores of poūdes. And farther you maye se, that the space betwene pennes and shyllynges may receaue but one counter (as all other spaces lyke wayes do) and that one standeth in that place for 6 d. Lyke wayes betwene the shyllynges & the poūdes, one coūter standeth for 10 s̄. And betwene the poundes and 20 li. one counter standeth for 10 poūdes. But besyde those you maye see at the left syde of shyllynges, that one counter standeth alone, & betokeneth 5 s̄.

So

Accomptynge

So agaynste the poundes, that one couter standeth for 5 li. And agaynst the 20 poundes, the one counter standeth for 5 score poūdes, that is 100 li. so that euery syde counter is 5 tymes so moch as one of them agaynst whiche he standeth. Now for the accompt of auditors take this example.

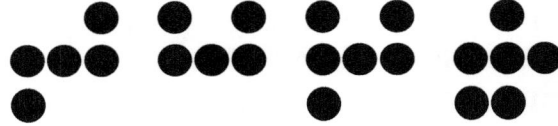

where I haue expressed ẏ same sūme 198 li. 19 s̄. 11 d̄. But here you se the pēnes stande toward ẏ ryght hande and the other encreasynge orderly towarde the lefte hande. Agayne you maye se, that auditours wyll make 2 lynes (yea and more) for pennes, shyllynges, & all other valewes, yf theyr summes extende therto. Also you se, that they set one counter at the ryght ende of eche rowe, whiche so set there standeth for 5 of that roume : and on the

by counters.

the lefte corner of the rowe it stãdeth for 10, of ỹ same row. But now yf you wold adde other subtracte after any of bothe those sortes, yf you marke ỹ order of ỹ other feate which I taught you, you may easely do the same here without moch teachynge: for in Addi tiõ you must fyrst set downe one sume and to the same set the other orderly, and lyke maner yf you haue many: but in Subtraction you muste sette downe fyrste the greatest summe, and from it must you abate that other eue ry denominatiõ from his dewe place. S. I do not doubte but with a lytell practise I shall attayne these bothe: but how shall I multiply and diuide after these fourmes? M. You can not duely do none of both by these sortes, therfore in suche case, you must resort to your other artes. S. Syr, yet I se not by these sortes how to expresse hũ= dreddes, yf they excede one hundred, nother yet thousandes. M. They that vse such accomptes that it excede 200

L.   in

Accomptynge

in one summe, they sette no 5 at the lefte hande of the scores of poundes, but they set all the hundredes in an other farther rowe, & 500 at the lefte hand therof, and the thousandes they set in a farther rowe yet, & at the lefte syde therof they sette the 5000, and in the space ouer they set the 10000, and in a hygher rowe 20000, whiche all I haue expressed in this example,

which is 97869 li 12 s. 9 d. ob. q̃. for I had not told you before where, nother how you shuld set downe farthynges, which (as you se here) must be set in a voyde space sydelynge beneth the pennes : for q̃ one counter : for ob. 2 counters : for ob.q̃. 3 counters : & more there can not be, for 4 farthynges

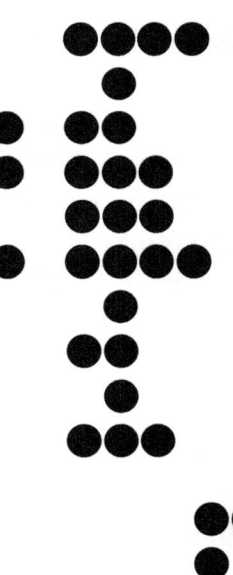

by counters.

do make 1 d. which must be set in his dewe place. And yf you desyre ẏ same summe after audytors maner, lo here it is.

But in this thyng, you shal take this for suffycyent, and the reste you shall obserue as you maye se by the wor=
kyng of eche sorte: for the dyuers wit=
tes of men haue inuented dyuers and sundry wayes almost vnnumerable. But one feate I shall teache you, whi
che not only for the straungenes and secretnes is moche pleasaunt, but al=
so for the good cōmoditie of it ryghte worthy to be well marked. This feate hath ben vsed aboue 2000 yeares at the leaste, and yet was it neuer comē=
ly knowen, especyally in Englysshe it was neuer taughte yet. This is the arte of nombrynge on the hand, with diuers gestures of the fyngers, expres
synge any summe conceaued in the
      L.ii. mynde

The arte of nōbrynge

mynde. And fyrst to begynne, yf you wyll expresse any summe vnder 100, you shall expresse it with your lefte hande: and from 100 vnto 10000, you shall expresse it with your ryght hande, as here orderly by this table folowynge you may perceaue.

⁋ Here foloweth the table of the arte of the hande.

by the hanbe.

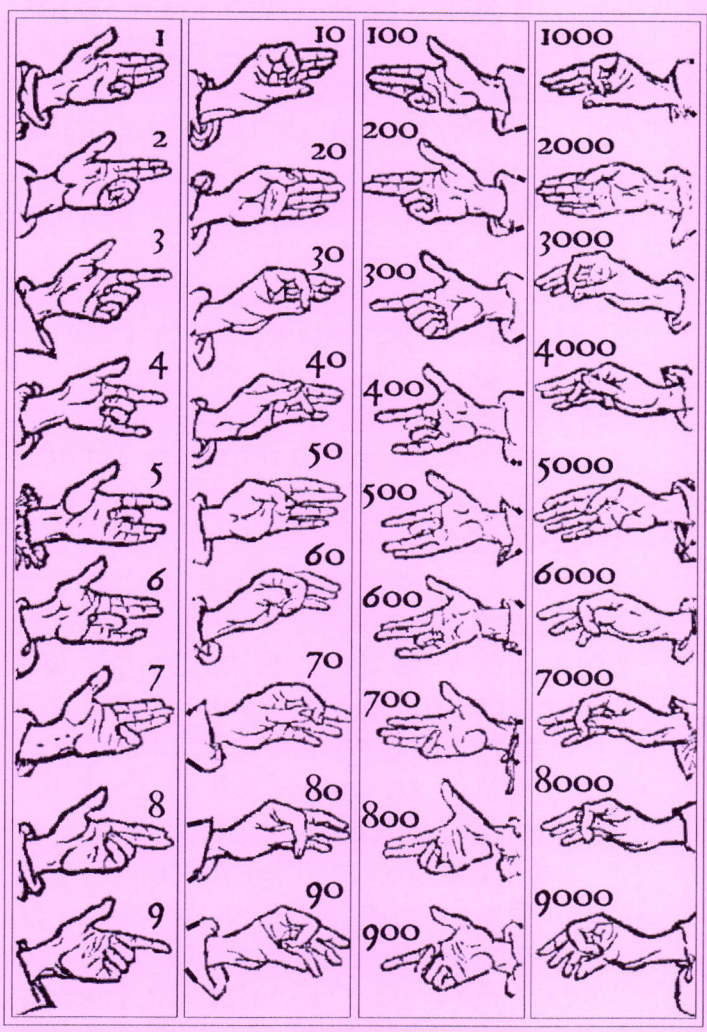

The arte of nōbrynge

1. In which as you may se 1 is expres=
sed by ẏ lyttle fynger of ẏ lefte hande
closely and harde crooked.

2. 2 is declared by the lyke bowynge of
the weddynge fynger (whiche is the
nexte to the lyttell fynger) together
with the lyttell fynger.

3. 3 is signified by the myddle fynger
bowed in lyke maner, with those o=
ther two.

4. 4 is declared by the bowyng of the
myddle fynger and the rynge fynger,
or weddynge fynger, with the other
all stretched forth.

5. 5 is represententh by the myddle fyn
ger onely bowed.

6. And 6 by the weddyng fynger on=
ly crooked: and this you may marke
in these a certayne order. But now 7,
8, and 9, are expressessed w̄ the bow=
ynge of the same fyngers as are 1, 2,
and 3, but after an other fourme.

7. For 7 is declared by the bowynge
of the lyttell fynger, as is 1, saue that
for 1 the fynger is clasped in, harde &
round

by the hande.

rounde, but for to expresse 7, you shall bowe the myddle ioynte of the lyttell fynger only, and holde the other ioyntes streyght. S. Yf you wyll geue me leue to expresse it after my rude maner, thus I vnderstand your meanyng: that 1 is expressed by crookynge in the lyttell fynger lyke the head of a byſſhoppes bagle: and 7 is declared by the same fynger bowed lyke a gybbet. M. So I perceaue, you vnderstande it.

Then to expresse 8, you shall bowe 8 after the same maner bothe the lyttell fynger and the rynge fynger.

And yf you bowe lyke wayes with 9 them the myddle fynger, then doth it betoken 9.

Now to expresse 10, you shall bowe 10 your fore fynger rounde, and set the ende of it on the hyghest ioynte of the thombe.

And for to expresse 20, you must set 20 your fyngers streyght, and the ende of your thombe to the partitiō of the

T.iiii.   fore

The arte of nōbrynge

foremoste and myddle fynger.

30 30 is represented by the ioynynge together of ẏ̆ headdes of the foremost fynger and the thombe.

40 40 is declared by settynge of the thombe crossewayes on the foremost fynger.

50 50 is sygnified by ryght stretchyng forth of the fyngers ioyntly, and ap=plyenge of the thombes ende to the partition of the myddle fynger & the rynge fynger, or weddynge fynger.

60 60 is formed by bendynge of the thombe croked and crossynge it with the fore fynger.

70 70 is expressed by the bowynge of the foremost fynger, and settynge the ende of the thombe betwene the 2 foremost or hyghest ioyntes of it.

80 80 is expressed by settynge of the foremoste fynger crossewayes on the thombe, so that 80 dyffereth thus frō 40, that for 80 the forefynger is set crosse on the thombe, and for 40 the thombe is set crosse ouer ẏ̆ forefinger.

90 is

by the hande.

90 is signified, by bendynge the 90
foze fynger, and settyng the ende of it
in the innermost ioynte of y̌ thombe,
that is euen at the foote of it. And
thus are all the nõbers ended vnder
100. S. In dede these be all the nom=
bers frõ 1 to 10, & then all the tenthes
within 100, but this teacyed me not
howe to expresse 11, 12, 13, &c. 21, 22, 23,   11, 12, 13,
&c. and suche lyke. M. You can lytell   21, 22, 23
vnderstande, yf you can not do that
without teachynge : what is 11? is it
not 10 and 1? then expresse 10 as you
were taught, and 1 also, and that is
11 : and for 12 expresse 10 and 2 : for
23 set 20 and 3 : and so for 68 you
muste make 60 and thereto 8 : and
so of all other sortes.

But now yf you wolde represente
100 other any nomber aboue it, you
muste do that with the ryghte hande,
after this maner.

You must expresse 100 in the ryght   100
hand, with the lytell fynger so bowed
as you dyd expresse 1 in the left hand.

T.b.

The arte of nōbrynge

200   And as you expressed 2 in the lefte hande, the same fasshyon in the ryght hande doth declare 200.

300   The fourme of 3 in the ryght hand standeth for 300.

400   The fourme of 4, for 400.

500   Lykewayes the fourme of 5, for 500.

600   The fourme of 6, for 600. And to be shorte: loke how you dyd expresse single vnities and tenthes in the lefte hande, so must you expresse vnities & tenthes of hundredes, in the ryghte hande. S. I vnderstande you thus:

900   that yf I wold represent 900, I must so fourme the fyngers of my ryghte hande, as I shuld do in my left hand to expresse 9, And as in my lefte hand I expressed 10, so in my ryght hande

1000   must I expresse 1000.

4000   And so the fourme of euery tenthe in the lefte hande serueth to expresse lyke nōber of thousādes, so ỹ fourme of 40 standeth for 4000.

8000   The fourme of 80 for 8000.

                                                     And

by the hande.

And the fourme of 90 (whiche is 9000 the greateſt) foꝛ 9000, and aboue that I can not expꝛeſſe any nomber. M. No not with one fynger : how be it, w dyuers fyngers you maye expꝛeſſe 9999, and all at one tyme, and that lacketh but 1 of 10000. So that vnder 10000 you may by your fyngers expꝛeſſe any ſumme. And this ſhall ſuſſyce foꝛ Numeration on the fyngers. And as foꝛ Addition, Subtraction, Multiplycatiō, and Diuiſion (which yet were neuer taught by any man as farre as I do knowe) I wyll enſtruct you after the treatyſe of fractions.

 And now foꝛ this tyme fare well,
  and loke that you ceaſe not to
  pꝛactyſe that you haue lear=
  ned. S. Syꝛ, with moſte
  harty mynde I thanke
  you, bothe foꝛ your
  good learnyng, &
  alſo your good
  couſel, which
(god wyllyng) I truſte to folow.
   Finis.

⸿ No hede so hedely can be geuen,
But errour slypperly wyll crepe in.
For man wout errour scarsely can be,
So ẏ errour excedeth all dyligencye.
Paciently therfore I praye you bere,
Those fewe fautes cōmytted here.
More pleasaūt pfyt I geue by reding
Thē greuous grefe by errours offen
                                           (dyng.

Lefe, syde, lyne, Correction.

Lefe	syde	lyne	Correction
2	2	14,	the more to be desyred.
5	1	4	I doubt not but this.
5	2	22	I wolde neuer blynne.
7	1	6	to cōtēde for the nōber.
7	2	15	places, wherin they be set
8	1	19	that place is laste, ẏ.
28	2	18	write beneth ẏ nether=moste.
32	1		mende bothe examples in theyr seconde summe thus, 6, 7, 3,
33	1	18	of ẏ shyllynges, which.
35	1	14	I rebate, and there resteth 9, which I &c.
41	1	16	I wyll propounde.
46	1	19	9400.

¶ Imprynted at London
in Powls church yarde
at the sygne of the
Brasen serpent
by R. Wolfe.
In the yeare of our Lord Christ
M. D. xliii.
in October

Printed in Great Britain
by Amazon